滨海地区
COASTAL AREAS
PROCEEDINGS OF COST-EFFECTIVE MEASURES FOR FLOOD CONTROL AND GROUNDWATER RECHARGE

防洪与地下水回灌的有效措施

张保祥 [德]W. F. 盖格 刘青勇／主编

中国环境科学出版社·北京

图书在版编目（CIP）数据

滨海地区防洪与地下水回灌的有效措施/张保祥，（德）盖格（Geiger，W），刘青勇主编. —北京：中国环境科学出版社，2012.5
ISBN 978-7-5111-0779-4

Ⅰ. ①滨… Ⅱ. ①张… ②盖… ③刘… Ⅲ. ①滨海—防洪—研究 ②滨海—地下水—水循环—研究 Ⅳ. ①TV877 ②P641.25

中国版本图书馆 CIP 数据核字（2011）第 242042 号

责任编辑　周艳萍
责任校对　唐丽虹
封面设计　彭　杉

出版发行	中国环境科学出版社
	（100062　北京东城区广渠门内大街 16 号）
	网　　址　http://www.cesp.com.cn
	联系电话：010-67112765（总编室）
	010-67112738（图书出版中心）
	发行热线：010-67125803，010-67113405（传真）
印　刷	北京市联华印刷厂
经　销	各地新华书店
版　次	2012 年 5 月第 1 版
印　次	2012 年 5 月第 1 次印刷
开　本	787×1092　1/16
印　张	15.25
字　数	350 千字
定　价	40.00 元

【版权所有。未经许可请勿翻印、转载，侵权必究】
如有缺页、破损、倒装等印装质量问题，请寄回本社更换

滨海地区防洪与地下水回灌的有效措施
编 委 会

主　编　张保祥　[德] W. F. 盖格　刘青勇

编　委（以姓氏笔画为序）

　　　　万　力　　王好芳　　王明森　　王维平　　王增亮　　刘建强

　　　　负汝安　　何茂强　　余　成　　吴泉源　　张　欣　　李传奇

　　　　李新军　　李福林　　陈学群　　孟凡海　　林洪孝　　范明元

　　　　赵亭月　　赵然杭　　赵德三　　曹　彬　　黄继文　　谭海鸥

　　　　G. 乌兹伯格　　P. 梅耶　　S.O. 卡登

前 言

水资源短缺和水污染问题在中国北方地区十分突出，严重影响着社会经济的发展以及人民生活水平提高和环境质量改善。特别是山东省深受缺水问题的困扰。在山东沿海地区由于海水入侵，降低了水资源的可利用性，使得缺水现象更加突出。严峻的水问题形势及所引起的社会经济危害迫使山东沿海地区采取多种措施方法，以缓解水资源短缺、减轻海水入侵危害。然而，事实证明，只有对这些措施进行有效地调控才能实现上述目标。在这种情况下，需要进行有效的水资源管理即综合可持续的水资源管理。

德国联邦教育与研究部（BMBF）和中国科学技术部（MOST）共同资助开展了题为"滨海地区防洪与地下水回灌的有效措施"的预研究项目，以评估该地区的水资源状况。该项目由德国杜伊斯堡—埃森大学城市水管理学院、德国水资源规划与系统研究所（WASY）、施莱格工程咨询有限公司联合山东省水利科学研究院、龙口市水务局、山东大学、山东农业大学及山东师范大学共同合作开展。他们曾为研讨会的召开及后续项目"山东滨海地区水资源综合管理技术"的申报作出贡献，该项目将于2011年12月结题。本书是预研究项目历次研讨会内容及部分合作研究成果的总结。

感谢张保祥研究员基于中德双方成员的研究成果对"滨海地区防洪与地下水回灌的有效措施"项目论文集的整理与汇编。

联合国教科文组织（UNESCO）可持续水管理委员会主席
W. F. 盖格
慕尼黑/济南
2010年12月

Preface

Water scarcity and water pollution are severe problems in the Northern part of China, seriously affecting socio-economic development and standards of living and environment. Especially Shandong province is plagued by water shortage. In the coastal catchments of the Shandong province water scarceness increased due to salt water intrusion, reducing the usability of available water resources. The pressing water problems in the coastal catchments in the Shandong province and resulting socio-economic dilemmas forced to implement a variety of measures to relieve water shortage and abate salt water intrusion. However, it proved, that little can be achieved, unless the measures are coordinated effectively. Such a situation calls for good water management, namely integrated and sustainable water resources management.

On the topic "Cost-Effective Measures for Flood Control and Groundwater Recharge in Coastal Areas" a study for the assessment of the situation was granted by the German Federal Ministry of Education and Research (BMBF) and the Chinese Ministry of Science and Technology (MOST). This project has been carried out by the Institute of Urban Water Management at the University of Duisburg-Essen, WASY GmbH and Schlegel GmbH & Co. KG together with Water Resources Research Institute of Shandong Province, Longkou Water Authority Bureau, Shandong University, Shandong Agricultural University and Shandong Normal University. They all contributed to workshops and formulated follow-up proposal on "Overall-effective measures for sustainable water resources management in the coastal area of Shandong Province, PR China", which again was granted by BMBF and MOST. This project will be completed by December 2011. This report summarizes the contributions to several workshops and findings of the fruitful cooperation.

These Proceedings on "Cost-Effective Measures for Flood Control and Groundwater Recharge in Coastal Areas" gratefully were compiled by Professor Zhang Baoxiang based on the contributions of German and Chinese partners.

UNESCO chair in Sustainable Water Management

W. F. Geiger

Munich/Jinan, December 2010

目 录

龙口市的水供求问题 .... 孟凡海 1
水污染防治在行动 .... 王增亮 6
龙口市的水资源问题和利用 .... 李传奇 12
基于 Virtual GIS 技术的龙口市流域综合治理研究 .... 吴泉源 17
基于熵权与 GIS 耦合的 DRASTIC 地下水脆弱性模糊
    优选评价 .... 张保祥 万 力 余 成 等 22
沿海地区水资源的合理配置与社会经济协调发展研究
    ——以山东省龙口市为例 .... 王好芳 31
山东半岛地下水库建设与研究进展 .... 刘青勇 张保祥 张 欣 等 35
海水入侵防治研究与实践进展 .... 李福林 赵德三 陈学群 42
莱州湾海（咸）水入侵区宏观经济水资源多目标决策
    分析模型 .... 何茂强 王维平 黄继文 等 49
流域生态需水——洪水的重要性 .... 贠汝安 57
山东省污水处理现状概述 .... 赵亭月 62
山东省农业用水价格改革方案及供水协会建设规划研究 .... 刘建强 68
水管理的社会经济作用分析 .... 谭海鸥 林洪孝 79
水资源陆海空协同系统研究 .... 赵然杭 90
莱州湾滨海流域农业灌溉用水对社会经济的影响及
    对策 .... 曹 彬 王维平 范明元 等 99
山东省滨海地区缺水造成的社会经济问题 .... 林洪孝 103
王屋水库饮用水水源污染风险性评价 .... 李新军 张保祥 孟凡海 107
城市水资源管理——全球性挑战 .... W. F. 盖格 113
可持续水管理——从目标到行动 .... W. F. 盖格 119
防洪与地下水回灌——中德合作北京示范项目 .... W. F. 盖格 132
从工程角度看成本效益 .... W. F. 盖格 P. 梅耶 143
环境保护措施的融资方式探讨 .... G. 乌兹伯格 152
WBalMo——水资源规划模型简介与应用实例 .... S. O. 卡登 157
Urban Water Management – A Global Challenge .... W. F. Geiger 166

Sustainable Water Management – From objective to
    implementation .... W. F. Geiger   174
Flood Control and Groundwater Recharge – Joint Chinese-German Demonstration
    Project for Beijing .... W. F. Geiger   192
Engineering View of Cost Efficiency .... W. F. Geiger   P. Meyer   205
Financing Possibilities for Environmental Protection Measures .... G. Würzberg   217
WBalMo: Water Resources Planning Model – Introduction and Practical
    Experience .... S. O. Kaden   225

龙口市的水供求问题

孟凡海
山东龙口市水务局，龙口

1　水资源状况

龙口市位于山东半岛北部，滨临渤海，陆地面积 893.32 km²，基本以山的分水岭为界，南部为低山丘陵，北部为平原，山丘区与平原区面积约各占一半。龙口市属暖温带半湿润季风型大陆性气候，四季分明，春季干燥多南风，夏季湿热多阴雨，秋季凉爽雨水少，冬季寒冷北风多。年平均气温 11.8℃，年平均蒸发量 1 238.2～1 350.2 mm，相对湿度 69%。

现状总人口 62.6 万人，城区人口 18 万人，为小城市。耕地面积 3 万 hm²，灌溉面积 2.7 万 hm²，粮食产量 3.2 亿 kg，果品产量 2.8 亿 kg，水产品产量 1.9 亿 kg，国内生产总值 123.8 亿元，人均 GDP 19 900 元。

龙口市多年平均降水量 586.3 mm，降水总量 5.24 亿 m³，从招远、栖霞、蓬莱入境水量 0.76 亿 m³，扣除地面蒸发、水体蒸发、植物蒸腾、排泄入海量外形成的水资源只有 2.3 亿 m³，人均占有水资源量 368 m³，为全国人均占有量的 16.7%。按联合国统计划分，人均水资源量小于 1 000 m³ 为缺水区，人均水资源量小于 500 m³ 为水危机区，龙口为水危机区，属资源型缺水。由于人口密度大，水资源量少，经济社会的发展对水的需求量与水资源的供给量之间的矛盾日益突出，建设节水型的社会是本地区发展的必然趋势。

本地区降水年际间变化大，20%频率年降水量 6.37 亿 m³，75%频率年降水量 4.28 亿 m³；枯水年连续时间长，20 世纪 80 年代后期连续 4 年干旱，90 年代后期连续 3 年干旱；年内分布不均，72.9%降水集中在 6—9 月份，用水量较大的 3—5 月份降水量占全年的 12.6%，常常是春旱、夏涝、晚秋又旱。

境内水系多发源于南部山区，向西北流入渤海，共有大小 23 条河流，主要河流有黄水河、泳汶河、曲栾河、龙口河、北马河、八里沙河，均为季节性河流。

龙口行政区域为近乎独立的水文地质单元。水资源量中可利用量为 1.63 亿 m³，其中地表水可利用量 0.80 亿 m³，地下水可利用量 0.83 亿 m³。按典型年计算水资源量，50%频率年水资源可利用量 1.61 亿 m³，75%频率年水资源可利用量 1.37 亿 m³，95%频率年水资源可利用量 0.90 亿 m³。地表水主要分布在南部山区，有大中型水库、小（一）型水库及小（二）型水库、塘坝，占地表水总量的 57%，地下水主要分布在北部平原区，仅黄水河流域下游和北马、中村一带的地下水量就占据了地下水总量的 51%。

2 水的供给

由于水源分布的差异，决定了采取的供水形式主要有两种：一种是集中供水形式，另一种是分散供水形式。城区和乡镇驻地多数采用集中供水形式，农村主要采用分散供水形式。

1958年以来，进行了大量的水资源开发利用工程建设，针对本地区地表水建有大中型水库3座，小型水库、塘坝354座，大型自控翻板拦河闸7座，水库干、支渠总长411km，固定机电排灌站86处，总库容2.23亿m^3，兴利库容1.23亿m^3；针对地下水建有地下水库2座，大口井573眼，各类机电井7548眼。水资源的开发利用程度比较高，已达到38%。

龙口市现状供水主干网络为"两纵三横"，两纵为王屋水库向北海经济园供水管线，长30km；北邢家水库（含迟家沟水库）向南山集团供水管线，长9km。三横为联合灌区供水明渠，干渠长80.2km；自来水供水管线，主管线长40km；百年电力供水管线，主管长25km。城区自来水供水能力达到14.5万m^3。

按水源划分，地表水供水量5000万m^3，地下水供水量12000万m^3/a，图1是1991—2003年实际供水量统计。

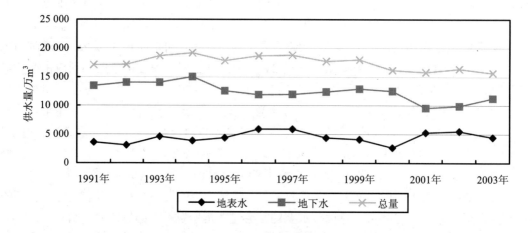

图1 龙口市的供水量统计

主要供水水源地有5处：王屋水库水源地、北邢家水库水源地、迟家沟水库水源地、莫家水源地、大堡水源地，其中生活供水水源地三处。王屋水库水源地为地表水源，建有两条供水管线，供水能力为4万m^3/d；北邢家水库水源地和迟家沟水库水源地为地表水源，各建一条供水管线，供水能力为1万m^3/d；莫家水源地在黄水河流域中游，为地下水源，供水能力为2万m^3/d；大堡水源地在黄水河流域下游地下水库范围内，供水能力为3.5万m^3/d。

按供水规模分，供水规模超过1万m^3/d的供水有五处，分别是向城区供水、向南山集团供水、向北海经济园区供水、向龙口市玉龙纸业有限公司供水、向山东百年电力发展股份有限公司供水。供水规模小于1万m^3/d的供水包括农村供水有500多个。

按供水目标分，城区生活供水530万m^3/a，占3%；农村生活供水1070万m^3/a，占

7%；工业供水 3700 万 m³/a，占 23%；农业供水 10900 万 m³/a，占 66%；环境供水 200 万 m³/a，占 1%。

现状大多数供水采用自来水供水形式，供水人口 57.24 万人，还有 5.36 万人采用自行取水形式，占总人口的 8.6%，主要分布在石良、黄山馆、北马、东江、芦头、七甲、下丁家、兰高、诸由观共 9 个镇。

现状集中供水的价格，地表水源供水原水价格：农业 0.09 元/m³，工业和城区 0.27 元/m³。城区公共供水价格分 4 种：居民生活 1.40 元/m³，公用事业 1.60 元/m³，工业 2.36 元/m³，饮食业 3.36 元/m³。

3 水的需求

按照现有人口和工业产业结构、规模，城区生活、工业、农业的需水情况为，现状城区生活需水量 530 万 m³/a；城区环境需水 50%保证率 80 万 m³/a、75%保证率 95 万 m³/a、95%保证率 95 万 m³/a；工业用水量 3700 万 m³/a；农村人畜需水量 1224 万 m³/a；粮田 50%保证率需水量 2400 万 m³/a、75%保证率需水量 2800 万 m³/a、95%保证率需水量 2800 万 m³/a；商品菜田 50%保证率需水量 2600 万 m³/a、75%保证率需水量 2880 万 m³/a、95%保证率需水量 2880 万 m³/a；林果 50%保证率需水量 5600 万 m³/a、75%保证率需水量 6720 万 m³/a、95%保证率需水量 6720 万 m³/a。合计 50%保证率需水量 16134 万 m³/a，75%保证率需水量 17950 万 m³/a，95%保证率需水量 17950 万 m³/a。

随着经济的快速发展，各行业对水的需求大量增加，到 2010 年，城区生活需水量为 1259 万 m³/a；城区环境需水量 50%保证率 189 万 m³/a、75%保证率 227 万 m³/a、95%保证率 227 万 m³/a；工业需水量 5400 万 m³/a；农村人畜需水量 994 万 m³/a；粮田 50%保证率需水量 2070 万 m³/a、75%保证率需水量 2430 万 m³/a、95%保证率需水量 2430 万 m³/a；商品菜田 50%保证率需水量 2560 万 m³/a、75%保证率需水量 2880 万 m³/a、95%保证率需水量 2880 万 m³/a；林果 50%保证率需水量 5100 万 m³/a、75%保证率需水量 6900 万 m³/a、95%保证率需水量 6900 万 m³/a。合计 50%保证率需水量 17572 万 m³/a，75%保证率需水量 20090 万 m³/a，95%保证率需水量 20090 万 m³/a。

4 供求存在的问题

4.1 资源短缺，浪费严重

龙口市人均水资源量 368 m³，为全国人均占有量的 16.7%，属资源型缺水地区，在这样一个缺水地区，人们珍惜水的意识还比较淡薄，浪费水的现象仍比较普遍，居民生活、公共设施、工业、农业不同程度上都存在着浪费水的情况。农业生产灌溉用水利用率 45%，比发达国家的 70%～80%低很多，工业用水的重复利用率不到 52%，与发达国家的 75%～85%有很大差距。不合理的用水结构浪费了大量水资源，生活、工业、农业、生态用水比例现状为 9.2∶16.1∶73.6∶1.1，农业用水仍占有很大的比重，农业节水仍将是今后的工作重点，工业的产业结构需要调整，减少耗水量大的企业。

4.2 供需矛盾突出

龙口水资源的开发程度较高,总体供水能力不会再有较大的增加,而经济发展的结果是需水量不断增加,这种情况会导致需水增加与供水不足之间的矛盾更加突出,预计2010年缺水将达到3 700万 m^3。

4.3 水资源过度开发已经导致水生态环境恶化

龙口市水资源过度开发利用已经引发了一系列的生态问题,最为突出的包括两方面:一是地下水超采引发的海水入侵,造成地下淡水咸化面积 103 km^2,向内陆纵向入侵最长达 4 km,3 000万 m^3 淡水失去利用价值,引发了许多生态、环境和工程问题;二是工业企业排放的大量污废水,污染了地下水源,造成水质性缺水。

4.4 水的宏观价格调控体系尚未建立

由于现状供水价格不高,不足以引起人们对节水的重视。现状水的价格体系需要政府重新核定,通过经济杠杆来减少对水的需求量。

5 解决的途径

5.1 调整产业结构,降低取水总量

农业是用水大户,应重点进行种植结构调整。深入实施农业种植业内部结构调整,减少高耗水作物的种植面积,增加经济作物的种植比例,发展高效节水农业。

工业积极发展高科技产业,根据龙口市国民经济和社会发展的实际情况,以新材料、生物技术和机电一体化三大领域为重点创新领域,带动电子信息技术、节能与环保技术和农业高新技术产业的开发和拓展,使新型建材、食品、电子电器、汽车关键零部件等四大支柱产业加速膨胀,快速发展,促进经济结构优化升级。高耗水、高污染、效益差的工业企业将被淘汰,新上的工业项目会受到严格的用水审批,要采用最先进的节水工艺。通过这些措施来降低取水总量。

5.2 提高供水价格,减少用水需求

新的水资源费标准已经实施,要切实落到实处,按照国务院和省政府的要求,尽快提高水利工程供水价格和城市用水价格,适度调整农业水价。实行用水定额管理、超定额累进加价制度。居民生活用水实行阶梯式计量水价,工农业生产用水根据用水定额和用水计划,实行定额用水平价、超定额超计划用水累进加价,农业用水实行库水预分、按亩配水、按方计费、预售水票、凭票供水、节约归己、超用累进加价的办法,促进节约用水。这些措施得到落实后,经测算,用水需求会下降10%。

5.3 推行节水措施,增加重复利用

工业企业与发达国家同行业相比,用水效益还不够高,应实施节水措施,提高工业用

水重复利用率，最终达到全部企业都有节水设备，水的重复利用率达到发达国家水平。目前，工业用水量占总用水量的12%，重复利用率为52%，低于国外先进水平75%，通过对现有工业冷却水和锅炉用水进行技术改造，增加生产工艺中的循环利用，工业取用水可节水25%。

5.4 处理排放污水，搞好中水回用

对于城区排放的污水，采取集中处理的方式，在东城区，经过两期施工，已建成日处理能力2.5万t的污水处理厂一座，处理后污水达到国家一级排放标准；在西城区，规划了龙港和龙口经济开发区两座污水处理厂，已经立项，即将实施。同时研究制定处理后的中水的回收利用的政策和措施。

对于工业企业，依法通过行政手段，促进工业企业建设和完善污水处理设施。目前，比较大的企业如南山集团、龙口电厂等均已建成了污水处理厂，日处理能力达到3.5万t，处理后的中水大部分在企业内部得到了利用，大大减少了污水的排放量。

对于农业，大力发展生态农业，推广节水灌溉和生物防治技术，使用高效、无污染的绿色肥料，推广生物农药，控制农药、化肥、地膜等污染，逐步削减化肥施用量，减少农业面源污染。

5.5 加强深度开发，提高雨洪利用

龙口水资源的开发程度虽然较高，但对于汛期过于集中的洪水无法得到有效利用，大部分洪水都排入渤海。对多余洪水的开发利用，应进行深入研究，加大开发力度，使洪水变成有效的资源。

5.6 增加海水利用，获取淡化水源

龙口是一个沿海小城市，有着丰富的海水资源，应充分加以利用。龙口电厂循环冷却水已使用海水，年用量约20000万m^3，还要扩大海水使用量，对能够直接利用海水的相关行业要进行技术改造，使之全部使用海水，这是一个不小的使用量，能够替代很多淡水资源。沿海地区，尤其是城区和海水入侵区，海水淡化将成为供水的一个重要水源，目前海水淡化成本还比较高，每立方米水5元多，推广有一定的难度，随着科技的不断进步，社会经济的发展，在不久的将来，海水淡化将在本地区得以实施。

水污染防治在行动

王增亮
山东省龙口市水务局，龙口

1 龙口概况

1.1 自然地理

龙口市地处山东半岛西北部，东经120°13′14″～120°44′46″，北纬37°27′30″～37°47′24″，陆地总面积893 km²，呈枫叶状，西部、北部濒临渤海，海岸线总长68.38 km。

龙口市总的地形是东南高、西北低，南部为低山丘陵，北部为平原，山区丘陵与平原面积约各占一半。龙口市属暖温带半湿润季风型大陆性气候，四季分明，季风进退明显。年平均气温11.8℃，多年平均降水量586.3 mm。

1.2 资源、能源

龙口自然资源丰富，境内建有全国唯一的低海拔大型海滨煤炭基地，褐煤总储量29亿t，年开采量500多万t；南部山区盛产黄金、花岗岩、石灰石等矿产资源；沿海大陆架石油及天然气储藏丰富。

龙口的城市功能日臻完善，交通便利，公路密度为每百平方公里为67.8 km；境内的龙口港为全国最大的地方港口，拥有万吨级以上泊位8个，年吞吐量超过1000万t。拥有全国第一座由中央和地方集资兴建的坑口电厂，全市目前各类电厂的总装机容量为160万kW。基础设施较为完备，城市化水平达到48%。

1.3 社会经济

龙口市辖14个镇（区、街），634个村（居），人口62万。近年来，龙口经济发展迅速，2003年完成国内生产总值192.9亿元，人均3.1万元。在全国经济竞争实力百强县中排名第25位，在山东省排名第三。

2 水资源与水工程

2.1 水资源特点

龙口市的水资源分布和特点主要是：总量少，多年平均降水量 5.24 亿 m^3，入境客水 0.76 亿 m^3，扣除地面蒸发、水体蒸发、植物蒸腾、排泄入海量，形成的水资源只有 2.3 亿 m^3，人均占有水资源量 368 m^3，为全国人均占有量的 16.7%。亩均水资源量 503 m^3；区域分布不均，57% 的地表水主要集中在南部山区，71% 的地下水主要分布在北部平原区；年际间变化大，丰枯具有连续性。龙口市多年水资源总量如表 1 所示。

表 1 龙口市水资源总量

典型年	地表水资源量/万 m^3	地下水资源量/万 m^3	重复计算量/万 m^3	水资源总量/万 m^3
多年平均	12 660	9 280	4 380	17 560
50%	12 009	9 280	4 250	17 039
75%	5 991	9 280	1 600	13 671
95%	3 398	9 280	1 200	11 478

2.2 水资源开发利用情况

地表水源工程。新中国成立后经过几十年的努力，到目前，龙口市已建有大型水库 1 座，中型水库 2 座，小（一）型水库 6 座，小（二）型水库 59 座，谷坊 600 座。在主河流——黄水河上建有大型钢筋混凝土翻板拦河闸 7 座。全市地表总蓄水能力达到 2.23 亿 m^3，总兴利库容 1.23 亿 m^3。

地下水源工程。全市目前有机电井 7548 眼，大口井 573 眼。在黄水河下游建有亚洲最大的地下水库，总库容 5359 万 m^3，最大调节库容 3929 万 m^3；在八里沙河中游建有小型地下水库一座，兴利库容 35.5 万 m^3。

2.3 水利工程在调配水资源方面的作用

通过修建水利工程，使水资源的使用和调配得到了有效的控制，地表拦水工程在汛期可以起到削减洪峰、减少洪涝灾害的作用，在汛末尽可能多地蓄水，为旱季浇灌及补充下游地下水储备充足水量；地下水利工程则在汛期内尽可能多地蓄水、回灌补源补充地下水，为旱季储备水量。各类拦蓄和调水、补源工程的建设，使水资源在时间和空间上得到了合理的调配。

2.4 良性循环的成功模式

黄水河流域总流域面积为 1034 km^2，在龙口境内的面积为 453 km^2，是龙口市工农业生产的主要水源地。按照"上游建地表水库，中游层层拦蓄、回灌，下游建地下水库，实行地表水、地下水联合调度；通过沿河修建污水处理厂和集中排放管道来保护水质"的建设思路，对黄水河主河道进行治理，使流域内水资源的"资源水、环境水、生态水及灾害水"的功能得到合理的转化和应用。目前,龙口市在黄水河主河道上游建成总库容为 1.21 亿 m^3 的大型地表水库 1 座,大型拦河闸 7 座,开挖各类补源渗井 2500 眼,年可复蓄水量 1942 万 m^3;下游建成了地下水库,库区最大回水面积 51.8 km^2,总库容 5359 万 m^3,最大调节库容 3929 万 m^3;在右岸堤外建设集中排污管道 9.01 km,建成日处理能力 4 万 t 污水处理厂 1 处;并对河道的河床、河堤等进行了治理。现在的黄水河河内碧波荡漾、两岸绿树成荫、鸟语花香,正在逐步变成一条真正的资源河、景观河和生态河,真正地成为龙口市人民的母亲河。

3 日新月异的经济发展与日渐凸显的水环境污染

3.1 产业比重的变化和工业结构变化

龙口市国民经济经过"八五"期间的快速发展和"九五"期间向市场经济的过渡发展,综合实力明显增强。国内生产总值及三次产业结构见表 2。

表 2 龙口市分期国内生产总值及三次产业结构比例

年份	国内生产总值/亿元	三次产业结构比例
1980	2.6	48.9:30.2:20.9
1990	14.4	19.1:60.8:20.1
1995	65.9	17.9:52.3:29.8
2000	123.8	12.9:56.1:31
2003	192.9	8.4:54.1:37.5

随着经济的发展,国民经济各行业中耗能大、耗水多的企业比重也在逐步加大。2003 年,作为龙口支柱产业的能源、铝业、机械、纺织、食品、建材及化工等行业的销售收入、利润分别占到了全市工业经济总量的 65% 和 68.5%。而这些行业大多是能耗大、耗水多、污染也较为严重。

3.2 用水量增加和用水比重变化

随着工农业的迅猛发展,各行业、各部门总的用水量逐步增加,而伴随着一些节水措施的实施,各行业的用水比重也随之变化。龙口市历年各部门用水量如表 3 所示。

表 3 龙口市历年各部门用水量统计

年份	农业用水/万 m^3	占总用水量的比例/%	工业用水/万 m^3	占总用水量的比例/%	生活用水/万 m^3	占总用水量的比例/%	总用水量/万 m^3
1991	14 966	85	1 555	9	1 116	6	17 637
1995	14 603	80	2 094	12	1 470	8	18 167
1999	14 003	77	2 664	15	1 528	8	18 195
2003	11 890	76	2 150	14	1 630	10	15 670

3.3 水污染源大致分布和基本量

2003 年，龙口市投入了大量的人力物力，对全市主要河流污染情况进行了调查，调查结果表明，共有沿河排污源 105 处，年向河道排放工业废水和生活污水总量约 680 万 m^3。污染源性质主要分为三大类：一是有机物污染。龙口市现有 31 家果脯厂、鱼片厂、粉丝厂、酿酒厂和 20 家造纸厂直接向河道排污，部分未经处理的城镇生活污水、医院和浴池污水等也直接进入河道。二是化学污染。此类污染源主要为色染厂、电镀厂、电解铝厂、雨布厂、制革厂、制药厂及选矿等，共有 17 家。另外，部分造纸厂废水也可造成一定的化学污染。三是悬浮物污染。目前，龙口市拥有众多的石材加工企业，其生产过程中有大量粉尘随废水进入河道，沉淀在河床表面，影响河道环境。此类排污企业共有 32 家。

3.4 重点污染行业及对当地的影响

重点污染行业排放的污染物主要为重金属和类金属污染物、酸碱及盐污染物、耗氧有机污染物等几类，这些污染物均可使水的溶解力、侵蚀性和化学活动性大大增强，破坏了水环境的化学稳定性，使水质的生物性功能恶化，硬度和矿化度的增高使水的工业用途受到极大的限制，这些污染物向河道内的排放，将使水资源的利用率降低，使当地水资源更加紧张。

3.5 对水环境的影响

水污染在河流中形成了自上游到下游的由分散到集中、由单一到多样的污染。污染源通过地表径流进入河道内，造成河道上游的水质恶化并下渗，污染后的地表水和沿河的浅层地下水无法使用，人们只能通过加大对深层地下水的开采来满足生产、生活需求，从而造成地下水位下降，形成地下水位负值区，引发污染水继续下渗，沿海地区发生海水入侵等结果，进一步加剧了水资源的短缺、深层地下水的污染程度及沿海区的水质咸化。水污染破坏了自然界的水均衡状态，造成了水环境的恶性循环。

4 觉醒的人们与不懈的努力

4.1 社会行动

水污染的问题已引起了龙口市人民政府的高度重视，设立了专门机构专项治理水污染。加大了污水治理的资金投入力度，建立了日处理能力为 4 万 t 的污水处理厂和污水集中排放管道系统。划定了地下水禁采区和地下水限采区，对全市的水资源进行动态管理和科学调度，并实行了用水企业取水许可制度，使有限的水资源得到合理调配。同时，加大宣传力度，引起全社会的关注并激发人们自觉保护水环境的意识。通过不定期的大力宣传，特别是近年来，龙口市为创建国家环保模范城市而进行的大规模宣传工作，使人们的环保意识普遍提高。主要表现在：一是大部分企业安装使用了节水设施，废、污水处理及回收利用设施，南山集团还自发兴建了日处理能力 3 万 t 的污水处理厂，最大可能地处理、回用废、污水，从而实现节水和达标排放，减少了对水环境的污染；二是在新建的居民住宅及宾馆、饭店、高层建筑等公用建筑中大量使用了先进的节水型卫生洁具、新型输水管材，部分老建筑也进行了改造，达到了节水标准。

4.2 治理规划

（1）地表治理和工程措施。按照"上游地表进行植树造林，修建水土保持工程，拦蓄水工程；中游河道层层拦截，修建回渗补源网络工程；下游平原区修建地下水库进行防止海水入侵、增加拦蓄量"的治理方案，对适宜修建水利工程的区域进行工程建设。同时，加大农业节水工程的建设力度，大力发展喷灌、微灌等高标准节水工程，在实现农业节水的同时，通过减少农药、化肥的施用量和深层渗漏量而减少对水环境的污染。

（2）污水处理厂建设。目前龙口已建有日处理能力4万t和3万t的污水处理厂各一处，规划再建设一处10万t/d的处理厂，并将4万t/d的处理厂扩建至10万t/d，同时尽量加长污水管道的延伸长度，并向支流方向延伸，严格控制污水排放。与此同时，通过舆论宣传和行政措施，强制一些较为分散的企业单独进行污水处理，使其做到达标排放。对分布松散的乡镇企业排放的经过处理且已达标的污水，通过工程措施蓄集起来用于农田灌溉，对污水处理厂处理的中水，可以输送到对水质要求不高的企业使用。

（3）节约用水的理念和措施。首先要加强水资源的管理，逐步实现依法治水和科学用水。尽快建立起高效有序的水资源管理体制，实现水资源管理一体化，同时要提高科学管理水平，对水资源进行长远规划，尽快构建水资源环境管理决策支持系统。还要转变管理理念，变资源水为商品水，收取与水资源价值相当的水费，利用价格手段消除因水价过低而导致的水资源浪费现象。要充分利用政策导向和经济杠杆，优化配置水资源，使有限的水资源持续产生出最大的经济、社会和生态环境效益。

5 本区域奋斗的目标

坚持科学发展观，多从本区域的实际出发，遵循自然规律和市场经济规律，以改善水环境、提高人民生活质量，实现可持续发展为目标，处理好长远与当前、全局与局部的关系，促进经济效益和生态效益的协调统一。具体措施如下。

5.1 唤醒大众意识，加强宣传

通过各种形式的宣传，如通过电视、广播、报纸等媒体宣传，宣传车下乡、发放宣传单、明白纸及树立宣传牌、环保提示牌等多种形式的宣传，特别是对广大的农村人群在发展绿色环保农业、减少无机肥料的使用和发展节水农业、减少灌溉用水量等方面的大力宣传，充分唤起大众的环保意识和水忧患意识，养成并保持保护环境、节约用水的生产和生活习惯。

5.2 合理设点布控，加强监测

在重点地区和主要水源地要根据每个监测点控制的面积，合理地布设监测点，设置观测点和观测井，布置试验观测仪器和设备，对水质好坏、水量多少和水位升降进行长期、实时、动态监测，以便及早发现问题、及早查出原因、及早进行治理。

5.3 科学运用现有工程治理模式，发挥示范推进作用

对现有的治理模式进行归纳、总结，分析其成功的经验和做法，并结合本区域的实际，实行计划用水和水资源的优化调度工作，加强示范、推广工作，使全社会都能参与到保护水资源的行列中来。

5.4 发挥工程集中治污与面上生态自然修复双向功能

在处理水污染的过程中，不仅要注重工程措施的落实，采取集中收集排放企业废水和污水、增大污水处理厂的处理能力、增加中水的回收使用量等方式加大循环水的利用率，还应注重生态修复的功能。通过采取植树造林、增大绿地面积及发展绿色农业、实现水土保持等措施，大量减少对水质的污染。同时，通过涵养水源、保持水土，有效增强土壤本身的生态恢复能力，起到生态修复、治理的作用和效果。

龙口市的水资源问题和利用

李传奇
山东大学土建与水利学院，济南

龙口市位于山东半岛沿海地区，地理位置优越，经济发达。由于水资源量少且年内、年际及地区间分布不均衡；工农业和社会发展对水的需求量增加；水资源的浪费和污染问题得不到根本的解决；水资源短缺的矛盾更加充分地暴露出来，水危机的状况日趋加重，水资源已成为龙口市经济社会可持续发展的"瓶颈"制约因素。

1 龙口市的水资源

水资源评价一般考虑地表水和地下水两部分，即在目前科技条件下能够开发利用的水量。

1.1 境内地表水资源量

地表水资源量指河流、湖泊等水体的动态水量，通常以河川径流量作为定量的指标，而大气降水是地表水和地下水的主要补给来源，在一定程度上能反映水资源量的丰枯情况，因此在水资源量评价中降水量的分析也是很必要的。

龙口市地域西、北临海，东、南部基本以山的分水岭为界，地表水资源量由天然入境客水量和境内自产径流量组成。龙口市境内多年自产径流量 7 093 万 m^3，天然入境客水量 5 576 万 m^3，天然径流总量 12 660 万 m^3。

1.2 地下水资源量

地下水资源量是指与降水、地表水体有直接补排关系的动态水量，主要是指矿化度小于 2 g/L 的淡水资源量。

龙口市地下水主要来自于降水入渗补给，其次为侧向补给，主要以河谷潜流形式进行补给。表 1 是龙口市多年地下水资源量计算结果。

综合各类型地区地表和地下的水资源量，扣除固有水力联系而重复部分便是本地区的水资源总量。把上述计算的地表水资源量和地下水资源量合计起来，为龙口市水资源总量（见表 2）。

表 1 龙口市地下水资源量

分区名称	年平均水资源量/万 m^3
南部山丘区	1 683
东部双灌区	1 770
东部井灌区	2 409
黄城区	450
龙口开发区	230
西部平原区	2 738
合计	9 280

表 2 龙口市水资源总量

典型年	地表水资源量/万 m^3	地下水资源量/万 m^3	重复计算量/万 m^3	水资源总量/万 m^3
多年平均	12 660	9 280	4 380	17 560
50%	12 009	9 280	4 250	17 039
75%	5 991	9 280	1 600	13 671
95%	3 398	9 280	1 200	11 478

1.3 龙口市水资源可利用量

由于受自然、技术、经济条件的限制以及环境对水的要求,水资源开发利用有一定的限度,既不可能,也不应该全部被利用。

水资源可利用量是指在经济合理、技术可能和生态环境许可的前提下,最大可能被人们控制利用的不重复一次性水量,包括地表和地下水资源的可利用量,但不包括河道内的用水量。它表示现状条件下,人类对水资源的调控能力。龙口行政区域为近乎独立的水文地质单元,水资源量中可利用量见表3。

表 3 龙口市水资源可利用总量

典型年	水资源总量/万 m^3	可利用总量/万 m^3	可利用率/%
多年平均	17 560	16 300	92.8
50%	17 039	16 100	94.5
95%	11 478	9 000	78.4

表3给出了龙口市水资源的承载能力,即在目前条件下,龙口市多年平均水资源可利用量为1.63亿 m^3。

2 水资源的特点及其利用中存在的问题

由于龙口市位于半湿润半干旱地带,蒸发潜力往往大于降水量。因此其水资源的显著特点是总量不足,人均、亩均水量偏低,水资源供需矛盾十分突出,对国民经济发展和社会生活的影响非常明显。

2.1 降雨量少，但蒸发量大

龙口市多年平均降水量 586.3 mm，年平均蒸发量 1 238.2～1 350.2 mm，蒸发量远大于降雨量。

2.2 降雨量时空分布不均

降水量年内分配不均，72.9%集中在 6—9 月份，仅 7—8 月份降水量即占全年的 49.8%，而 3—5 月份降水量占全年的 12.6%，10—11 月份占全年的 9.3%。形成春旱、夏涝、晚秋又旱的气象特点。降水量年际变化大，1964 年全市平均降水量 1 046.2 mm，1989 年仅为 329.4 mm，极差 716.8 mm。降水地域分布不均，南部山区降水量较大，北部平原区降水量较小。

丰枯具有连续性，20 世纪 80 年代后期连续 4 年出现枯水年，90 年代中期连续 3 年出现丰水年，90 年代后期又连续 3 年出现枯水年。

2.3 人均占有水资源量少，供需矛盾突出

龙口市人均水资源占有量 368 m³，为全国人均占有量的 16.7%。亩均水资源量为 503 m³，为全国的 27.9%。按联合国统计资料划分，人均水资源量小于 1 000 m³ 为缺水区，小于 500 m³ 为水危机区，龙口为水危机区，属资源型缺水。龙口市人口密度大，水资源条件先天不足，是本区水资源供需矛盾突出的主要原因。在 2001 年的现状工程组合条件下，2010 年 50%保证率供水量 17 573 万 m³，需水量 21 469 万 m³，缺水 3 084 万 m³，缺水率 14.7%。

本区水资源不足，严重影响国民经济发展。

3 缺水问题的分析

3.1 缺水性质的分析

缺水与否是水资源问题中最为重要也是最为根本的问题。然而，由于水资源的短缺受社会、经济和自然等因素的综合影响，从不同的研究目的和观察角度出发，缺水的内涵也不尽相同。

3.1.1 缺水的三种定义

从气候学的角度定义缺水：根据区域水热平衡的观点，分析区域缺水情况。当一个地区的蒸发能力（PE）大于降水（P）时，就认为该区域出现了气候学上的缺水。即 PE>P，缺水量等于 PE-P，此时，进行生态环境建设一定要考虑生态需水量。如我国的北方地区，从气候学的角度看，都有不同程度的缺水问题，生态和社会经济用水的矛盾是比较普遍的。

从社会的角度定义缺水：当一个地区的社会用水量大于当地水资源量时，必然出现缺水。可分为三种类型，即设施型、管理型和水质型。

从经济学的角度定义缺水：以边际效益为标准，当水的开发利用投入的边际效益为正时，表明当地的水资源开发利用在经济发展上仍是合理的，需要增补水量，以满足经济效

益，而未予满足时，则认为是缺水状态。这类缺水可以从调整结构或适应性措施来解决。

3.1.2 对龙口市缺水性质的认识

按气候学上定义，龙口市是缺水的，因为 PE/P 的年值为 1.3~1.5。从社会经济上看，龙口市人口密度平均为 700，属社会性缺水，但仍可人为控制需水，进行产业结构调整等。从龙口市用水分类中可以看出，农业用水 14511 万 m^3，占总用水量的 81%，仍占有很大的比重。龙口市经济发展超过水资源的承载力。例如，由于地下水的超采和乱采，导致大面积海水入侵，已严重污染了该区西部及北部地下水环境。龙口市 1984 年海水入侵面积 64.5 km^2，1988 年为 78.1 km^2，四年平均每年入侵 3.47 km^2。

3.2 水资源变化和缺水分析

以上的缺水评价是以年平均计算的，实际上，由于降雨年际变化很大，使缺水分析变得很不确定。龙口市属于干旱和湿润的过渡区，总体上是半湿润，但冬春季，干旱很严重，当地对冬春降水敏感，冬春缺水成为经常状态。在气候变化的背景下，研究表明，水资源的减少是非线性的，降雨量的减少，加上蒸发增加，水资源的极值两极分化明显，冬春缺水，夏秋洪涝，因此，用年均值不能完全反映龙口市缺水的真实情况。

3.3 人类活动的影响巨大，生态环境用水被大量挤占

目前龙口市的地表水利用率已达 66%，地下水开采利用率高达 90%以上，有些地方已经超采。深层地下水补给缓慢，而超采尤为严重，已出现了一系列的环境问题。与此同时，由于工业和城市发展迅速，污水排放量增加，环境自净能力下降，水质严重恶化。

4 建议和思考

解决龙口市水资源短缺的途径，总的来说要做到开源、节流和保护三者并重，开源与节流使水资源供需平衡，水源保护是供需平衡的基本保证，三者关系处理的好坏牵涉到本地区的社会安定、经济繁荣、人口素质和环境保护等一系列重大问题。

4.1 合理安排生态环境用水

龙口市缺水是事实，如何解决？水资源数量和质量评价应从以上三种定义来考虑，气候学定义的缺水地区，在水资源供需平衡分析时，应充分考虑生态用水的需求，保证必要的生态用水量。

最小生态需水量是维系生态环境系统基本功能的一种水量。过去在经济社会用水较低的时代，人类与生态系统两种用水的矛盾并不突出，曾一度忽视了生态用水量。当经济社会用水不断加大，大到挤占必要的生态蓄水量时，生态失去平衡而恶化，这才引起人们的思考，才开始注意研究生态需水的深刻意义。龙口市属于干旱半湿润地区，生态环境脆弱，必须确保生态环境用水，防止灌溉农业的盲目发展和生态环境的进一步恶化。

4.2 供需分析，不应只是增加供水，而应从需水控制去考虑，同等研究供需两个方面

（1）需水：应遵循"节水优先，治污为本，多渠道开源"的原则，进行需水管理，有

效控制需水的盲目增长。

农业是国民经济的基础，用水量大、集中。目前龙口市农业用水占国民经济用水的 80%以上，因此农业节水十分重要。

节水的另一条途径是污水资源化。城市污水未经处理排放入河道，既浪费了资源，又污染了环境。据统计，1997 年，龙口市工业废水和城市生活污水排放量合计 1 863 万 m^3，将来随着用水量的增加，污水量还会增加。如对这些污水加以处理，达到环境允许的排放标准或污水灌溉的标准，使污水资源化，既可增加水源解决农业缺水问题，又可起到治理污染的作用。将污水处理回用于农业并与污染治理有机地结合起来，对于解决龙口市水资源的短缺和水环境的改善将有特别重要的意义。

（2）供水：应遵循"先当地，后调水"的原则，最大限度地提高当地现有水利设施的利用和调度潜力，加强当地水源和各种代用水源的合理开发，如处理后的污水、海水利用、微咸水利用、雨水利用等。龙口市有入渗条件很好的地面，并有大量的地下蓄水空间，可用来进行地下水人工回灌，而汛期的雨洪水和处理后的达标污水，将是人工回灌的主要来源。

（3）供需平衡：从供水量和需水量、污水利用、用水效率来分析，龙口市的水资源有效利用率较低，单方水的产出明显低于发达国家，节水尚有较大潜力。节约用水和科学用水，应成为水资源管理的首要任务；对经济型缺水，应提高经济用水的水平，调整产业结构，合理布局，减低单位产品的耗水量。缺水地区应以供定用，提高用水效率。

基于 Virtual GIS 技术的龙口市流域综合治理研究

吴泉源
山东师范大学人口、资源与环境学院，济南

1 引言

虚拟现实（Virtual Reality，VR）是一种新兴的计算机综合信息分析技术，它利用计算机生成逼真的三维视觉、听觉、触觉等形式的虚拟世界[1]。它是在计算机图形学、多媒体技术、人工智能、人工接口技术、高度并行的实时计算技术和人的行为学等一系列高新技术的基础上发展起来的一种技术，并且是这些技术在更高层次上的集成和渗透。GIS 是建立在数据库系统之上的一种管理事物空间关系的管理信息系统，是收集、存储、转换、分析、显示地理信息的系统，是进行区域、资源和环境规划管理的辅助决策手段，是信息技术、计算机技术发展与地理、测绘等地球科学相结合的产物[2]。RS 是采集地球数据及其变化信息的重要技术手段，它由地面信息特征、信息的获取、信息的传输与记录、信息的处理和信息的应用等五个部分组成[3]。将虚拟现实（VR）技术与地理信息系统（GIS）技术和遥感（RS）技术集成起来用于虚拟地理环境便构成了虚拟再现地理信息系统[4]（Virtual GIS，VR-GIS）。Virtual GIS 不仅具有 GIS 的各项功能，而且由于 Virtual GIS 所表示的空间信息与现实世界相似，视觉效果好，可以较容易地被人们所理解，因而应用起来比较方便，特别是近几年来由于高分辨遥感图像的出现，进一步促进了 Virtual GIS 技术的发展。因此，尽管 Virtual GIS 技术起步晚，但发展迅速，得到了广泛应用。将 Virtual GIS 技术用于流域环境研究，通过对不同时期流域地理环境的虚拟再现，从多空间、多角度来追踪流域地理环境的演化过程，认识其存在的问题，在此基础上经过三维建模分析，结合龙口市水资源时空分布特点[5, 6]及经济发展规律，提出科学合理的治理措施，达到治理效果先知的目的。

2 研究区概况

龙口市地处山东半岛北部，其西部、北部濒临渤海，南部与栖霞市和招远市毗邻，东部与蓬莱市接壤，陆地总面积 893.32 km^2；地势南高北低，在地质构造上位于鲁东断块之胶北块隆的西北部，地貌受黄县大断裂和玲珑—北沟断裂控制，北部为断陷盆地，南部为断隆山地；研究区属温带半湿润季风型大陆性气候，四季分明，季风进退明显，春季干燥多南风，夏季湿热多阴雨，秋季凉爽雨水少，冬季寒冷北风多，年平均气温 11.8℃，无霜期 160~210 天，年平均蒸发量 1295mm，相对湿度 69%，多年平均降水量 586mm，且主要集中在 6—9 月，降水变差系数 0.27；南部低山区，植物生长茂盛，主要为针阔叶混交林，中部丘陵区主要为经济林，北部平原区则是农业种植区，南北景观分异十分明显；受人类活动的强烈影响，龙口市的生态环境发生了较大变化，许多支流被改道，上游水库密布，下游渠道纵横。人均占有水资源 386m³，分别占山东省和全国人均占有量的 41.9%和 16.7%，属严重缺水地区。

3 流域 Virtual GIS 的构建

3.1 资料与数据处理

构建龙口市流域 Virtual GIS 需要多种空间数据，其中主要包括大比例尺地形图、多时相、多光谱和高分辨率的数字遥感图像。地形图采用的是 1996 年航测的 1∶10000 数字影像地形图，高斯—克吕格投影模式。为了能够反映研究区地理环境的动态变化，遥感图像分别采用了 1972 年航空相片、1992 年、1998 年、2000 年、2003 年 TM 和 ETM+卫星图像。数据处理包括高程数据的采集，统一坐标系，遥感图像的校正、增强以及多源栅格数据融合等项内容。

3.2 Virtual GIS 系统的构建

龙口市流域 Virtual GIS 的构建主要是在遥感图像处理软件 ERDAS 系统内完成的，包括 DEM 建立、地表信息集成、Virtual GIS 视景设置、Virtual GIS 分析、Virtual GIS 导航等项工序[7]，详见图 1。

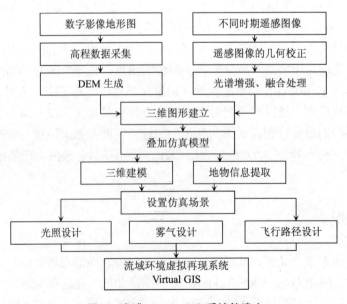

图 1 流域 Virtual GIS 系统的建立

4 应用结果分析

4.1 流域特征分析

（1）河流水系：龙口市境内有大小河流 23 条，主要河流有：黄水河、曲滦河、泳汶河、龙口北河、北马南河、八里沙河、界河等。其中黄水河是流经龙口市最大的河流，干流全长 55.43 km，流域面积 1 034.47 km²，在龙口市境内面积 452.97 km²。上述河流的流域形态特征基本相似，皆源于南部山区，属季节性河流，上游为低山丘陵区，河网密集，河流形态、走向严格受构造、地貌控制，多呈树枝状结构；下游为断陷盆地，河网稀疏，水系呈平形状排列。从再现的 1972 年流域状况看出，平原区的河道宽窄不一，曲率较大，而从再现的 2003 年流域状况，平原区的河道整齐，宽窄基本一致，说明人类活动对平原区的河流水系影响较大。

（2）水利工程：新中国成立以来，龙口市对其境内的河流水系进行了多次治理改造，在中上游修建了大型水库 1 座，中型水库 2 座，小（一）型水库 6 座，小（二）型水库 59 座，塘坝 289 座，拦河闸 4 座，谷坊 600 座，总库容 2.73 亿 m³，兴利库容 1.23 亿 m³，多年平均蓄水量 9 417.5 万 m³。地表水资源开发利用率达到 70% 以上，远远高于全国平均水平。下游平原河道经过多次治理疏通、裁弯取直，河道的防洪能力基本达到了 20 年一遇的标准。1990 年又在黄水河下游兴建了国内第一座地下水库，库容量达 2 986 万 m³，多年平均蓄水量 2 430 万 m³，对缓解下游地区的缺水问题和防治海水倒灌起到了积极作用。

4.2 存在的主要问题

（1）河道工程：在不同时期河网虚拟再现的基础上，经水文模拟分析，发现河道工程问题较多。一是河道淤积严重，河道内草岛面积近 30 年增加了 2.8 倍。草岛的大量存在，导致河道有效过水断面变小，过水能力下降，有些河段的防洪能力由原来的 20 年一遇，降低到目前的 10 年一遇。二是坝体单薄，部分河段坝体几乎全无，且河堤内、外边坡不足，部分河堤为砂质坝，无植被保护，极易冲刷破坏。三是河道内被大量取砂，加之洪水的冲刷，导致河床不断下降，部分河唇及浆砌防护段基本已掏空，危及堤坝安全。四是堤坝管理不善。河道保护范围内经常进行各种生产活动，破坏了河道的行洪安全。五是河道污染严重，部分河道垃圾成岭，污水横流。

（2）水土流失：再现不同时期流域地表环境发现流域内的土地覆盖变化非常大，尤其是在平原与山区之间的环境脆弱地带，林地、耕地、草地大面积缩小，从而导致了水土流失现象的加重。水土流失使水库淤积、水域萎缩、河床抬高。而且水土流失引起土地贫瘠和土地沙化，使生态环境恶化，导致水旱灾害增多，农业收入下降，严重影响农业经济的发展。

（3）地表蓄水工程：首先是水库重叠现象严重，有些水库的蓄水能力是其所在流域多年平均地表水资源量的 1.5～2.0 倍，既造成了经济损失，又占用了大量土地。其次是蓄水工程存在着质量问题。由于地表蓄水工程大都兴建于 20 世纪五六十年代，受当时经济、技术条件的限制，水坝质量较差，一旦出现问题，破坏性较大，经行洪模拟分析，可淹没土地 6.8 万亩，淹没村庄 79 个。

4.3 流域综合治理措施

4.3.1 工程措施

（1）坡面治理

山顶：在大于 20°的山顶，主要措施是封山育林，稀疏林补植，使林草保存面积达到宜林草面积的 80%以上。

山腰：在 15°～20°的山腰，大力发展各类干杂果、水果等经济林，加强山腰旱薄地的改造。通过采取深翻平整土地，将"三跑田"变为"三保田"。坡面要建好环山公路，路面要有排水设施，防止造成水土流失，并加强管护。

山脚：15°以下的山脚要整成水平梯田或"三合一"梯田。

（2）沟道治理

从流域上游毛、支沟到中下游骨干河道和入海口，按设计的防洪标准进行整治，建设各类拦蓄工程，最大限度地拦截泥沙，拦蓄地表水，提高水资源的利用率。

毛沟：按照小流域治理的要求，自上而下修建谷坊，形成谷坊群，使整个毛沟层层拦截，达到泥沙不出沟，地表水就近拦蓄。谷坊的防洪标准，按 5～10 年一遇 3～6 小时最大暴雨设计，谷坊在布局上要因害设防。在毛沟的上游，应以修建干砌石拦沙谷坊为主，拦截泥沙，稳定侵蚀基点；在毛沟的中下游，应以修建浆砌石蓄水谷坊和土谷坊为主，拦蓄地表水。谷坊的间距，原则上是下一道谷坊的顶部大致与上一道谷坊的底部等高。

支沟（小型河道）：主要指骨干河道的一、二级支流。按照 20 年一遇的防洪标准，搞好筑堤、修路、绿化等综合整治，使堤、路、林相互配套。在沟道内，根据不同地形、地质情况，因地制宜地修建人字闸、截流坝、迷宫堰等各类拦蓄工程，使整个沟道层层拦蓄，道道截流，形成工程相连，水面相映。各类拦蓄工程的断面尺寸应符合标准，做到安全稳定，经济实用。

骨干河道：黄水河、泳汶河、北马南河、八里沙河四条河道。按照防洪标准，搞好设计进行综合整治，达到河床平、堤成行、闸相连、路相通、树成行、渠配套的要求，充分发挥防涝排洪、水资源开发利用、环境绿化美化等综合效益。适宜建闸的河道，要建闸拦蓄。闸的形式要因地制宜，采取砌石坝、翻板闸、人字闸、迷宫堰等。黄水河综合治理工程就成功地创造了"上游修建地表水库地表拦蓄，中游设闸层层拦补，下游建地下水库地下拦截"的综合治理模式，每年为龙口市增加可利用水资源 3 000 多万 m^3。

（3）蓄水工程治理

对现有地表蓄水工程进行质量评估，加固拦水坝，增建引水工程，提高蓄水能力和水资源的利用率。对布局不合理的水利工程，该改建的改建，该拆的拆，要合理拦蓄地表水资源量。从全流域生态环境的整体出发，全盘考虑地表水资源拦蓄量，确保下游地区有足够的生态需水量。

4.3.2 生物措施

（1）营林造林，防治水土流失：森林被誉为"绿色水库"，在雨季，能把一部分降水蓄存起来，减少径流，减轻洪涝灾害；在旱季，被存储的水又缓慢地释放出来，增加径流量。因此，森林有调节流量，防御旱涝的作用。在配套实施工程措施的同时，大力发展水土保持林，种植刺槐、赤松、麻栎、杨树等树种，并积极引进优良品种，发展经济林，提高森林覆盖率。搞好绿化造林，沿堤坝两侧种植树木，是治理水土流失的基本措施。各河流流域的绿化造林搞好了，既能防止水土流失，又能减少灾害，促进流域社会经济的发展。

（2）调整种植结构，发展节水农业：还林还果，引进优良品种，实行林、粮、果合理间作。加大农业生产结构调整力度，调整耕地、林地、经济林的比例。在南部低山丘陵区，建设林果经济区，种植龙口长把梨等经济林木；北部平原地区，建设粮油、蔬菜和畜牧经济区；沿海建设葡萄、海产经济区。三区互相促进，共同发展，促进龙口经济快速发展。结合发展高效农业和菜篮子工程，大力推广管灌、喷灌、滴灌、渠道防渗等节水技术，逐步实现粮田管灌喷灌化、果园微灌管灌化、大棚微喷滴灌化、渠道浆砌防渗化、农田灌溉科学化、工程管理企业化，努力节省水资源，提高水的有效利用率。

4.4 治理效益预测分析

上述措施实现后将会达到五个方面的治理效果：一是防洪标准得到进一步提高，能够达到设计洪水标准，减淹面积 12.1 万亩，减淹人口 13.6 万人、村庄 114 个和主要公路 4 条，年防洪效益达 3 869 万元；二是地表、地下水资源的开发利用率明显提高，较大程度地解决龙口市水资源的供需矛盾。拦河闸等一系列拦蓄工程的建设，年增加可利用水量 669 万 m^3，加之原渗井补源工程的有效利用，年增加地下水入渗量 1 880 万 m^3、可利用水量 1 380 万 m^3，年增加工农业生产效益 3 360 万元；三是水土流失面积得到控制。治理水土流失面积 86 km^2，占全市应治理面积的 58%，年保水量 470 万 m^3，年减蚀量 31 万 t，年增效益 2 480 万元；四是水资源紧缺状况进一步得到缓解。地下水位比治理前上升 0.6～0.8 m，地下水位负值漏斗区面积由现有的 81.5 km^2 减少到 53 km^2，海水入浸得到控制，入浸面积有所减少；五是生态环境得到极大的改善，人民生活质量进一步提高，促进了全市工农业的发展。

参考文献

[1] 王雪梅，马明国，李新. VR-GIS 技术在数字黑河流域飞行模拟中的应用[J]. 遥感技术与应用，2002，12（6）.
[2] 汤国安，赵牡丹. 地理信息系统[M]. 北京：科学出版社，2000：1-8.
[3] 梅安新，彭望禄，秦其明，等. 遥感导论[M]. 北京：高等教育出版社，2001：1-3.
[4] 董浩. VR-GIS 技术在小流域治理规划中的应用研究[J]. 水利水电技术，2002，5（2）.
[5] 吴泉源. 龙口市水资源环境管理决策支持系统构建研究[J]. 地理科学，2001，10（5）.
[6] 吴泉源. 龙口市水资源时空分析信息系统构建研究[J]. 地域研究与开发，2003，8（4）.
[7] 党安荣，王晓东，陈晓峰，等. ERDAS IMAGINE 遥感图像处理方法[M]. 北京：清华大学出版社，2002：426-507.

基于熵权与GIS耦合的DRASTIC地下水脆弱性模糊优选评价

张保祥[1,2]　万　力[3]　余　成[2]　孟凡海[4]
1. 山东省水利科学研究院，济南；
2. 北京大学工学院水资源研究中心，北京；
3. 中国地质大学水资源与环境学院，北京；
4. 山东省龙口市水务局，龙口

地下水是国民经济发展的重要物质基础，在保障城乡居民生活、支撑社会经济可持续发展和维持生态平衡等方面发挥了重要作用。近年来，随着经济的快速发展和人类活动的影响，与地下水有关的环境问题日益凸显。目前，我国浅层地下水资源污染比较普遍，全国浅层地下水大约有50%的地区遭到不同程度的污染，约有一半城市市区的地下水污染比较严重，地下水水质呈下降趋势。

地下水保护是在地下水还没有或几乎没有受到污染的情况下，根据研究区的水文地质条件、水化学条件及污染物特性等情况圈划出污染敏感带，为管理人员提供决策依据。对地下水脆弱性分区评价是对含水层进行保护的主要手段之一。地下水脆弱性评价研究的对象主要是针对目前最普遍、最难以控制和治理的地下水非点源污染问题。通过对地下水的脆弱性进行评价，可以给出不同分区地下水的脆弱程度，评价地下水潜在的易污染性，圈定地下水的脆弱性范围，最终使人们在开发利用地下水资源的同时，采取有效的防治措施，以确保地下水资源的可持续利用。

地下水脆弱性是一个相对模糊概念，其评价的方法也有多种。目前，国际上应用最普遍、最成熟的地下水脆弱性评价方法是DRASTIC评价指标体系[1-3]。该方法在美国、加拿大、南非及欧共体各国成功应用并积累了相当丰富的经验，近年来在我国的应用也得到了迅速发展。

1　DRASTIC模型介绍

地下水脆弱性DRASTIC评价模型由美国水井协会（NWWA）和美国环境保护局（USEPA）在20世纪80年代提出[4]，它综合了40多位水文地质学专家的经验。他们认为，地下水遭受污染的风险大小取决于污染源和含水层本身所固有的水文地质特性——易污染性。对同一污染源，不同的含水层，由于气候、土壤、地形条件以及地质构造、水文地质条件的不同，即含水层的脆弱性程度不同，会造成不同的易污染性。由于地下水脆弱性取决于含水层本身固有的特性,因此该方法选取对含水层脆弱性影响最大的 7 项水文地质参数评价指标来定量分析地下水的脆弱性,即:D——地下水埋深(Depth to the Water);R——含水层净补给量(Net Recharge);A——含水层岩性(Aquifer Media);S——土壤类型(Soil Media);T——地形坡度(Topography);I——非饱和介质影响(Impact of the Vadose Zone Media);C——含水层水力传导系数(Hydraulic Conductivity of the Aquifer)。DRASTIC 即由上述 7 项指标的英文代表字母组成。它的应用假设条件如下:污染物由地表进入地下;污染物随降雨入渗到地下水中;污染物随水流动;评价区面积大于 40.5 hm^2。

DRASTIC 模型评价指数通常用数字大小来表示,它由 3 部分组成:权重、范围(类别)和评分,其意义如下。

(1)权重:每一个 DRASTIC 评价指标根据其对地下水防污性能的作用大小都被赋予一定的权重,权重值大小为 1~5,最重要的评价参数取 5,最不重要的评价参数取 1。各评价指标权重取值的大小要结合具体的评价区域来选定,DRASTIC 各评价指标的权重值的大小见表 1,权重为不可改变的定值。DRASTIC 权重的赋值分为两种情况,一种适用于自然条件下的地下水本质脆弱性评价;另一种是针对强烈的农业活动区设计的,也被称作农药权重,专门用于特殊脆弱行评价。

表 1 DRASTIC 模型中各评价指标及其权重(L. Aller 等,1987)

指标		D	R	A	S	T	I	C
权重	自然条件	5	4	3	2	1	5	3
	农药喷洒	5	4	3	5	3	4	2

(2)范围(类别):对于每一个 DRASTIC 评价指标来说,根据其对地下水防污性能的作用大小可以分为不同的范围(数值型指标)和类别(文字描述性指标),见表 2。

(3)评分:每个 DRASTIC 评价指标都可用评分值来量化这些数值范围和类别对地下水污染的可能影响,其评分取值范围为 1~10,分别对应于每一评价指标的变化范围(类别),见表 2。

表 2 地下水脆弱性 DRASTIC 评价指标的范围(类别)和评分(L. Aller 等,1987)

地下水位埋深		净补给量		含水层岩性			土壤类型		地形坡度		非饱和介质类型			水力传导系数	
范围/m	评分	范围/mm	评分	类别	评分	典型评分	类别	评分	范围/%	评分	类别	评分	典型评分	范围/(md^{-1})	评分
0~1.5	10	0~51	1	厚层页岩	1~3	2	薄层或无/砾石	10	<2	10	承压层	1	1	0.05~4.9	1
1.5~4.6	9	51~102	3	变质岩/火成岩	2~5	3	砂层	9	2~6	9	粉土/黏土	2~6	3	4.9~14.7	2
4.6~9.1	7	102~178	6	风化变质岩/火成岩	3~5	4	泥炭层	8	6~12	5	页岩	2~5	3	14.7~34.2	4
9.1~15.2	5	178~254	8	冰碛层	4~6	5	涨缩性黏土	7	12~18	3	变质岩/火成岩	2~8	4	34.2~48.9	6

地下水位埋深		净补给量		含水层岩性			土壤类型		地形坡度		非饱和介质类型			水力传导系数	
范围/m	评分	范围/mm	评分	类别	评分	典型评分	类别	评分	范围/%	评分	类别	评分	典型评分	范围/(m·d^{-1})	评分
15.2~22.9	3	>254	9	层状砂岩、灰岩及页岩序列	5~9	6	砂质壤土	6	>18	1	灰岩	2~7	6	48.9~97.8	8
22.9~30.5	2			厚层砂岩/厚层灰岩	4~9	6	壤土	5			砂岩/层状砂岩、灰岩及页岩/砂、砾石与粉土黏土互层	4~8	6	>97.8	10
>30.5	1			砂和砾石	4~9	8	粉砂壤土	4			砂及砾石	6~9	8		
				玄武岩	2~10	9	黏壤土	3			玄武岩	2~10	9		
				岩溶发育的灰岩	9~10	10	腐质土	2			岩溶发育的灰岩	8~10	10		
							非涨缩性黏土	1							

DRASTIC 地下水脆弱性评价指数为以上 7 项指标的加权总和，由下式确定：

$$D_{in} = \sum_{j=1}^{7}(w_j \times r_j) \tag{1}$$

式中：D_{in}——DRASTIC 指数；w_j——指标 j 的权重；r_j——指标 j 的评分。

根据计算出的 DRASTIC 评价指数，就能够识别地下水污染敏感区，具有较高脆弱性指数的区域的地下水就易于被污染，反之亦然。DRASTIC 评价指数提供的仅是相对概念，并不表示地下水污染的绝对数值。因此由正常和农药两种情况获得的 DRASTIC 评价指数并不等同。对于正常情况，DRASTIC 评价指数的最小值为 23，最大值为 226，而在农药喷洒情况下的最小值与最大值分别为 26 和 256。一般地下水脆弱性 DRASTIC 评价指数值在 50~200。

2 熵权模糊优选方法原理

熵的概念是 1865 年由德国物理学家克劳修斯（R. Clausius）首先提出的，最初属于热力学范畴，到目前为止，熵的应用已经涉及物理、数学、化学、生物、地理、信息及社会科学等领域。1948 年，美国工程师申农（C. E. Shannon）将熵引进到信息论研究中，他将信息源的熵定义为[5]：

$$H = -C\sum_{i=1}^{n} p_i \ln p_i \tag{2}$$

其中 $p_i \geq 0$，$\sum_{i=1}^{n} p_i = 1$；p_i（$i=1, 2, \cdots, n$）为信息源中第 i 种信号出现的概率；$-\ln p_i$ 为它带来的信息量；C 为比例系数。上式表明熵和状态的概率有密切关系，从而成为系统状态不确定程度的度量。

近 20 多年来，熵的应用已经被引入到模糊数学领域，构成模糊熵，用它从整体上描述系统的不确定性，即对模糊集所含模糊性大小的一种度量。在实际应用中，把多目标决策评价各待选方案的固有信息和决策者的经验判断的主观信息进行量化和综合，进而可以建立基于熵的多目标决策评价模型。其建模步骤如下[6~8]：

（1）设有 n 个待评价的样本（叠加分区），反映各分区脆弱性的评价指标有 m 个，则根据实测数据可构造评价指标特征值矩阵 X：

$$X_{ij} = \begin{bmatrix} x_{11} & x_{12} & \cdots & x_{1m} \\ x_{21} & x_{22} & \cdots & x_{2m} \\ \vdots & \vdots & \cdots & \vdots \\ x_{n1} & x_{n2} & \cdots & x_{nm} \end{bmatrix}, \quad (i=1, 2, \cdots, n;\ j=1, 2, \cdots, m) \quad (3)$$

式中，X_{ij} 为第 i 个样本的第 j 项指标的特征值。

（2）由于参与评价的地下水脆弱性 DRASTIC 模型中各项指标有越大越优型（D、T）和越小越优型（R、A、S、I、C），因此，需对矩阵（3）中的特征值进行归一化处理，方法如下：

$$\begin{cases} r_{ij} = \dfrac{x_{ij} - \min(x_j)}{\max(x_j) - \min(x_j)}; & \text{越大越优型} \\ r_{ij} = \dfrac{\max(x_j) - x_{ij}}{\max(x_j) - \min(x_j)}; & \text{越小越优型} \end{cases} \quad (4)$$

式中，r_{ij} 为第 i 个样本的第 j 项指标对脆弱度的相对隶属度；$\max(x_j)$、$\min(x_j)$ 为指标 j 中的最大值、最小值。对各指标进行归一化处理，得到相对隶属度矩阵 R：

$$R_{ij} = \begin{bmatrix} r_{11} & r_{12} & \cdots & r_{1m} \\ r_{21} & r_{22} & \cdots & r_{2m} \\ \vdots & \vdots & \cdots & \vdots \\ r_{n1} & r_{n2} & \cdots & r_{nm} \end{bmatrix} \quad (5)$$

（3）计算第 j 个评价指标下第 i 个待评价样本评价指标特征值比重：

$$p_{ij} = r_{ij} - \sum_{i=1}^{n} r_{ij} \quad (6)$$

（4）计算第 j 个评价指标的熵：

$$e_j = -\frac{1}{\ln(m)} \sum_{i=1}^{n} p_{ij} \ln p_{ij} \quad (7)$$

（5）计算第 j 个评价指标的权重：

$$w_j = 1 - e_j - \sum_{j=1}^{m}(1 - e_j) \qquad (8)$$

设第 i 个样本对脆弱性的相对隶属度为 u_i，m 个指标的权重向量为 $W = (w_1, w_2, w_3, w_4, \cdots, w_m)^T$，满足归一化条件，即 $\sum_{j=1}^{m} w_j = 1$。根据模糊集理论可将隶属度定义为权重[9-11]，则加权广义距离：

$$D(r_i) = u_i \sqrt{\sum_{j=1}^{m}\left(w_j |r_{ij} - 1|\right)^p} \qquad (9)$$

上式完善地描述了样本 i 距脆弱性的距离。为求解 u_i 的最优值，建立目标函数：

$$\min\left\{F(u_i) = u_i^2 \left[\sum_{j=1}^{m}\left(w_j|r_{ij}-1|\right)^p\right]^{\frac{2}{p}} + (1-u_i)^2 \left[\left(\sum_{j=1}^{m} w_j r_{ij}\right)^p\right]^{\frac{2}{p}}\right\} \qquad (10)$$

式中，p 为距离参数，p 值为 1 时为海明距离，p 值为 2 时为欧氏距离，这里 $p=1$。对 u_i 求偏导数并使之等于 0，得：

$$u_i = \left\{1 + \frac{\left[\sum_{j=1}^{m} w_j(1-r_{ij})\right]^2}{\left(\sum_{j=1}^{m} w_j r_{ij}\right)^2}\right\}^{-1} \qquad (11)$$

公式（11）即是可应用于若干叠加分区的地下水脆弱性评价模糊优选理论模型。

3 GIS 评价过程

地理信息系统（GIS）的基本技术是地理空间数据库、地图可视化和空间分析，它将地球表层信息按其特性的不同进行分层，每个图层存储特征相同或相似的事物对象集。它能够把地理位置和相关属性信息有机地结合起来，可有效地收集、存储、处理、分析和显示空间信息，具有地域综合、空间分析、动态预测与提供决策支持等功能[12]。目前，GIS 已被广泛应用于城市管理、土地评价、国土规划、地籍测量、环境监测及地质、气象、农业、林业、交通等部门。

空间叠加分析是 GIS 中重要的空间分析功能之一，它是将多幅图层相重叠生成新的图形单元和重新组合属性的过程。它将有关主题层组成的数据层面，进行叠加产生一个新数据层面的操作，其结果综合了原来两层或多层要素所具有的属性。叠加分析不仅包含空间关系的比较，还包含属性关系的比较。地下水脆弱性 DRASTIC 模型的评价过程就是将 DRASTIC 的 7 个指标分区图利用 GIS 软件进行空间叠加分析[13-16]，流程见图 1。通过叠加分析运算产生一个新的区文件，该文件是由 7 个参数图层文件在空间上的弧段进行求交运算形成的图斑文件，该文件的属性结构包含了 7 个参数图层文件所有的属性结构和内容。熵权模糊优选评价方法是在 DRASTIC 模型中各指标分区叠加结果基础上将各新分区的属性数据进行熵权模糊优选评价，然后将评价结果（隶属度）返回给分区属性值，再利用此值对地下水脆弱性各分区进行再分类和评价。上述过程是在 MapGIS 系统平台下，采用 Visual C++编程语言二次开发方式来完成的。

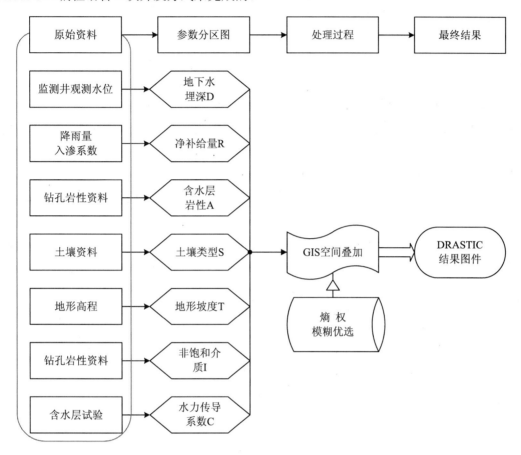

图 1 基于 GIS 的 DRASTIC 地下水脆弱性评价方法流程图（A. Rahman，2008，有改动）

4 应用实例

黄水河流域位于山东半岛北部，面积 1 034.47 km²，其中平原区约占 20%。黄水河地下水库位于黄水河中下游平原区，库区最大回水面积 51.8 km²，总库容 5 359 万 m³，最大调节库容 3 929 万 m³。库区北面濒临渤海，东西两侧为低山丘陵，地面高程 1.7～43 m。黄水河流域地下水以大气降水及河川径流入渗补给为主，第四系孔隙水为主要地下水类型，广泛分布于黄水河中下游冲洪积平原、山前倾斜平原、滨海堆积及黄水河故道中，是地下水主要开采区，含水层由 Q_4^{a1} 砾质粗砂、Q_4^m 粗、中、细砂、$Q_3^{pl\sim al}$ 砾质粗砂、$Q_2^{pl\sim al}$ 中粗砂组成[17]。选取黄水河中下游地区作为评价区，总面积为 319.97 km²，占整个黄水河流域总面积的 30.9%。

针对黄水河流域的具体情况，对研究区地下水脆弱性评价 DRASTIC 模型的各个指标分区分别进行数字化，并把每个指标的范围（分类）和评分值分别赋予其相应的属性参数中去进行叠加，共形成 93 个新分区。采用开发出的基于熵权与 GIS 耦合的 DRASTIC 地下水脆弱性模糊优选评价系统对模型所需的 7 个参数叠加后的分区结果进行计算，依据公式（7）和公式（8）计算出的各指标的熵和权重见表 3，利用公式（11）求得各叠加分区的熵权模糊优选结果（隶属度）见表 4。

表 3 DRASTIC 模型各指标的熵和权重

	D	R	A	S	T	I	C
熵	0.00673655	0.00641688	0.00615842	0.0060091	0.0068138	0.00659521	0.00637345
权重	0.142815	0.142861	0.142898	0.14292	0.142804	0.142835	0.142867

表 4 各叠加分区熵权模糊优选结果

分区 ID	1	2	3	4	5	6	7	8	9	10
优属度	0.7048	0.6399	0.4641	0.4998	0.6399	0.7641	0.3599	0.1595	0.1831	0.4287
分区 ID	11	12	13	14	15	16	17	18	19	20
优属度	0.8403	0.9436	0.9436	0.8765	0.9000	0.8622	0.7643	0.8237	0.9049	0.3268
分区 ID	21	22	23	24	25	26	27	28	29	30
优属度	0.5710	0.2357	0.1180	0.1378	0.2646	0.5356	0.3598	0.6058	0.3268	0.0999
分区 ID	31	32	33	34	35	36	37	38	39	40
优属度	0.1180	0.2085	0.8874	0.8001	0.2357	0.8621	0.9049	0.9401	0.2646	0.4288
分区 ID	41	42	43	44	45	46	47	48	49	50
优属度	0.8621	0.2645	0.8874	0.2357	0.1180	0.5356	0.2646	0.8874	0.4288	0.2950
分区 ID	51	52	53	54	55	56	57	58	59	60
优属度	0.1378	0.1378	0.2357	0.3598	0.3598	0.2645	0.9163	0.8621	0.7352	0.9000
分区 ID	61	62	63	64	65	66	67	68	69	70
优属度	0.3268	0.6730	0.8621	0.2646	0.4288	0.9000	0.3268	0.7352	0.0999	0.1180
分区 ID	71	72	73	74	75	76	77	78	79	80
优属度	0.8621	0.2949	0.3268	0.8765	0.0837	0.1378	0.9436	0.3988	0.9000	0.6731
分区 ID	81	82	83	84	85	86	87	88	89	90
优属度	0.7353	0.8621	0.6731	0.8621	0.8820	0.9206	0.9163	0.8468	0.8820	0.9049
分区 ID	91	92	93							
优属度	0.9163	0.7642	0.8622							

根据隶属度最大原则，结合黄水河流域的实际情况和地下水管理工作的需要，将地下水脆弱性分为 3 个等级：①高脆弱性，优属度为 0.0837～0.3599；②中脆弱性，优属度为 0.3600～0.7988；③低脆弱性，优属度为 0.7989～0.9436。相应的地下水脆弱性分区结果见图 2。

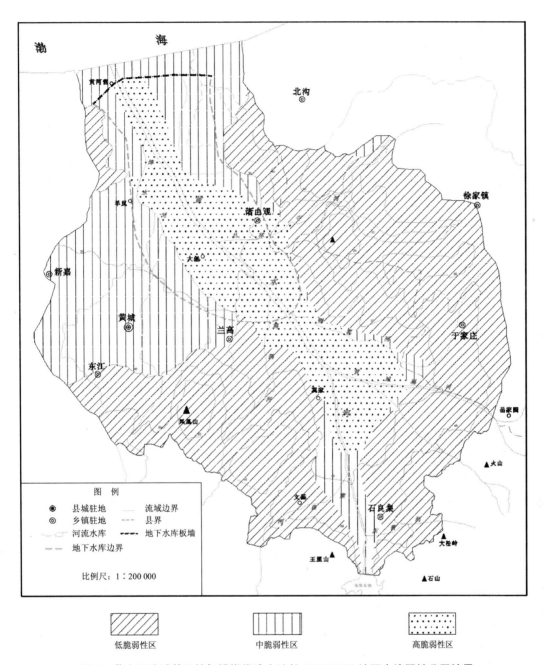

图 2 黄水河流域基于熵权模糊优选方法的 DRASTIC 地下水脆弱性分区结果

5 结语

　　针对传统的地下水脆弱性评价 DRASTIC 模型中需要人为确定指标权重的缺陷,本文将熵权系数法引进该模型,建立了基于熵权的地水脆弱性评价 DRASTIC 模糊优选评价模型,该模型避免了人工确定权重的主观性,使评价结果更加客观、可靠。

　　在基于 GIS 的地下水脆弱性评价基础上,与熵权模糊优选方法进行了耦合。通过二次开发，使得分析评价及运算结果可及时显示，克服了以往图件和数据相互分离的弊端。提升了结果的可靠性和预测的准确性，能为主管部门的分析决策提供强有力的依据。

利用上述方法对黄水河流域进行了地下水脆弱性评价，结果表明，与传统方法相比，该方法与实际情况更加吻合。

今后，GIS 技术与数学模型结合将是地下水脆弱性评价主要的发展方向和趋势。

参考文献

[1] 张丽君. 地下水脆弱性和风险性评价研究进展综述[J]. 水文地质工程地质，2006（6）：113-119.

[2] 钟佐燊. 地下水防污性能评价方法探讨[J]. 地学前缘，2005，12（增）：3-11.

[3] 李志萍，许可. 地下水脆弱性评价方法研究进展[J]. 人民黄河，2008，30（6）：52-54.

[4] Aller, L., T. Bennett, J. Lehr, et al. DRASTIC: A Standardized System for Evaluating Ground Water Pollution Potential Using Hydrogeologic Settings[R]. National Water Well Association PB87-213914，EPA 600/2～87/035，1987.

[5] 王彬. 谈熵[M]//新疆维吾尔自治区科学技术协会编. 熵与交叉科学[C]. 北京：气象出版社，1988：19-22.

[6] 姜蕊云，赵庆超，苗永宏. 基于熵权的 DRASTIC 模型在地下水脆弱性评价中的应用[J]. 黑龙江水专学报，2008，35（3）：63-65.

[7] 张少坤，付强，张少东，等. 基于 GIS 与熵权的 DRASCLP 模型在地下水脆弱性评价中的应用[J]. 水土保持研究，2008，15（4）：134-137.

[8] 刘仁涛，付强，李国良，等. 基于熵权的 DRASTIC 模型及其在地下水脆弱性评价中的应用[J]. 农业系统科学与综合研究，2007，23（1）：74-77.

[9] 王国利，周惠成，张文国. 含水层易污染性评价的模糊优选方法[J]. 水利学报，2000，31（12）：72-76.

[10] 陈守煜，王国利. 含水层脆弱性的模糊优选迭代评价模型及应用[J]. 大连理工大学学报，1999，39（6）：811-815.

[11] 陈守煜. 水资源与防洪系统可变模糊集理论与方法[M]. 大连：大连理工大学出版社，2005：241-250.

[12] 宫辉力，赵文吉，李小娟. 地下水地理信息系统——设计、开发与应用[M]. 北京：科学出版社，2006：1-8.

[13] 阮俊，肖兴平，郑宝锋，等. GIS 技术在地下水系统脆弱性编图示范中的应用[J]. 地理空间信息，2008，6（4）：55-57.

[14] Insaf S. Babiker, Mohamed A. A. Mohamed, Tetsuya Hiyama, et al. A GIS-based DRASTIC model for assessing aquifer vulnerability in Kakamigahara Heights, Gifu Prefecture, central Japan[J]. Science of the Total Environment. 2005，345：127-140.

[15] D. Thirumalaivasan, M. Karmegam, K. Venugopal. AHP-DRASTIC: software for specific aquifer vulnerability assessment using DRASTIC model and GIS[J]. Environmental Modelling & Software, 2003, 18：645-656.

[16] Atiqur Rahman. A GIS based DRASTIC model for assessing groundwater vulnerability in shallow aquifer in Aligarh, India[J]. Applied Geography, 2008, 28：32-53.

[17] 孟凡海，柳杨科，吴泉源，等. 水资源与龙口市可持续发展[J]. 山东师范大学学报（自然科学版），2002，17（2）：59-62.

沿海地区水资源的合理配置与社会经济协调发展研究
——以山东省龙口市为例

王好芳

山东大学土建与水利学院，济南

龙口市位于山东半岛沿海地区，总面积为 893 km²，地理位置优越，经济发达。与本区经济发展不相协调的是水资源严重短缺，水资源总量仅为 2.3 亿 m³，人均水资源占有量 368 m³，为全国人均占有量的 16.7%。由于水资源量少且年内、年际及地区间分布不均衡，工农业和社会发展对水资源的需求量增加，水资源短缺的矛盾更加加剧。水资源已成为龙口市经济社会可持续发展的制约因素。因此，为了更好地实现水资源的可持续利用与社会经济的协调发展必须进行水资源的合理配置。

1 水资源的合理配置原则

根据龙口市可利用水资源的特点和各用水户对水质、水量在不同时段的要求及用水效益状况进行水资源的合理配置，为此应遵循以下原则：

（1）动态性原则：不同时期区域内可利用的水资源是不同的，并且随着社会经济的发展，该区域的产业结构也会发生变化，各行业或用水部门对水量和水质的要求也会不同。因此，在配置过程中要根据不同时期各用水部门对水资源的不同要求进行动态的配置。

（2）时空性原则：各行业或各用水部门对水资源需求的时段不同，而且各用水部门或供水水源在区域内的空间位置也不同。所以，在水资源的合理配置过程中还要考虑时间上和空间上的合理性。

（3）协调性原则：由于水资源水量有限而且水质不同，各用水户对水资源的量和质的要求也是不同的，同时，需求时段也不同。为此，在水资源的合理配置过程中要协调水资源和各用水户之间的供需矛盾，使有限的不同水质的水资源，满足不同用户在不同时段的需求。

（4）效益性原则：水资源合理配置的效益不是区域内某一部门的效益最大，而是整个区域的社会、经济、生态环境综合效益最好。

2 水资源合理配置的目标

水资源的合理配置是针对水资源短缺和用水的竞争性提出来的[1]，目的是使有限的水资源能更好地满足社会经济的发展，根据上述配置原则，水资源合理配置的目标是使区域的经济、环境、社会协调地发展。具体来说就是通过在生活、工业、农业和生态环境各用水户之间的合理配置，使整个区域的用水净效益达到最大。

3 基于量与质的多目标水资源配置模型[2]

3.1 子模型形式

3.1.1 水量分析模型

水资源的合理配置是解决水资源的短缺和用水竞争性问题，更好地协调和满足各用水户的需求。从水量上来分析，水资源的合理配置应该使水资源的缺水量达到最小。模型形式为

$$\min \sum_t \sum_n (\sum_i Q_{int} - Q_{ntd})^2 \tag{1}$$

$$\text{s.t} \quad \begin{array}{l} Q_{int} \in Q_{int\,max} \\ \sum_i Q_{int} \leqslant Q_{ntd} \end{array}$$

式中：Q_{int} 为第 i 种水资源给 n 个用水户 t 时间内的供水量；Q_{ntd} 为 n 个用水户在 t 时间内的需水量；$Q_{int\,max}$ 为 t 时间内 i 种水资源所能提供的最大水量。

3.1.2 水质分析模型

为了使有限的水资源更为有效地发挥其使用价值，在其合理配置时应充分考虑水资源的水质，使不同水质的水资源尽量满足不同行业或用户的生产、生活需要。其模型形式为

$$\min \sum_t \sum_n (w_j \sum_i Q_{int} - w'_k Q_{ntd})^2 \tag{2}$$

$$\text{s.t} \quad k \in j$$

式中：w_j 为水质为 j 类的水资源在供水量中所占的比重；w'_k 为不同用水户对第 k 类水资源的需求比重；Q_{int}、Q_{ntd} 同上。

3.1.3 经济分析模型

水资源合理配置的目标之一就是为了取得更好的经济效益。因此，在解决了水资源的短缺和用水竞争问题之后，应使其经济效益最好，模型形式为

$$\max(\sum_{nt} B_n \sum_i Q_{int} - \sum_{int} C_{in} Q_{int}) \tag{3}$$

$$\text{s.t} \quad \begin{array}{l} Q_{int} \in Q_{int\,max} \\ \sum_i Q_{int} \leqslant Q_{ntd} \end{array}$$

式中：B_n 效益系数；C_{in} 为费用系数。

另外，在配置中也要考虑生态环境效益和社会效益。

3.2 协调模型形式

水资源的配置是一个复杂的大系统,在水资源的合理配置过程中要实现其中的一个目标是容易的,但是要同时实现所有的目标是很难的。根据上述的分析模型及前人的研究工作[3],充分考虑水资源对社会、经济和生态环境的支撑和制约,利用大系统理论和多目标决策理论,建立基于量与质的社会、经济和生态环境综合净效益最优的水资源配置协调模型,其形式为:

$$\max\left\{\sum_n\sum_i\sum_j (w_l Q^l_{ijnt} + w_f Q^f_{ijnt} + w_a Q^a_{ijnt} + w_e Q^e_{ijnt})\right\} \quad (4)$$

$$\text{s.t} \begin{cases} Q^l_{ijnt} + Q^f_{ijnt} + Q^a_{ijnt} + Q^e_{ijnt} \leq Q_{ijnts} \\ Q^l_{ijntd} + Q^f_{ijntd} + Q^a_{ijntd} + Q^e_{ijntd} = Q_{ijntd} \\ Q^l_{ijnt} \leq Q^l_{ijntd} \\ Q^f_{ijnt} \leq Q^f_{ijntd} \\ Q^a_{ijnt} \leq Q^a_{ijntd} \\ Q^e_{ijnt} \leq Q^e_{ijntd} \end{cases}$$

式中:Q^l_{ijnt}、Q^f_{ijnt}、Q^a_{ijnt}、Q^e_{ijnt}、Q_{ijnts} 分别为第 i 种水资源中水质为第 j 类的水资源给区域 n 个用水户 t 时段的城镇生活供水量、工业供水量、农业供水量、生态环境供水量和总供水量;w_l、w_f、w_a、w_e 分别为城镇生活供水量综合效益权重、工业供水量综合效益权重、农业供水量综合效益权重和生态环境供水量综合效益权重;Q^l_{ijntd}、Q^f_{ijntd}、Q^a_{ijntd}、Q^e_{ijntd}、Q_{ijntd} 分别为对第 i 种水资源中第 j 类别的水资源区域 n 个用水户 t 时段的城镇生活需水量、工业需水量、农业需水量、生态环境需水量和总需水量。

上述公式(1)~(4)中 $i=1, 2, 3, \cdots, m$;$j=1, 2, \cdots, n$;m 为水资源种类总数,n 为水质的总类数。

4 建立龙口市水资源合理配置决策支持系统

4.1 系统总体功能框架

龙口市水资源合理配置决策支持系统的主要功能是:在不同水平年条件下,计算典型年及长系列水资源供需平衡,在上述配置原则、目标及配置模型支撑下,进行龙口市水资源的合理配置,为决策者提供各种决策信息。具体框架见图 1。

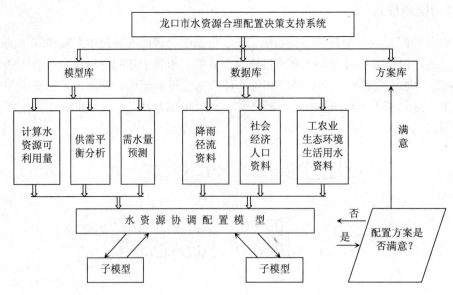

图 1　水资源合理配置决策支持系统框架图

4.2　系统运行模式

该决策系统包括三个库：模型库、数据库和方案库。其中数据库是三库中的基础，它是计算、求解所需数据的来源。模型库是手段，系统只有通过对模型的求解才能给出决策需要的信息。方案库是核心，它是模型运行结果的表现，并对决策者提供有力的支持。三库之间的关系是：数据库为模型库提供模型运行所需要的数据；模型运行的结果以一定的存贮格式存入数据库。模型库对数据库提出数据要求和存贮格式，同时模型库又是方案库的支持者。方案库是数据库和模型库最终的运行结果。

5　结论

水资源的合理配置是一个复杂的大系统，它涉及社会经济系统和自然生态系统。通过建立水资源合理配置决策支持系统能够对龙口市水资源的持续利用提供有效支撑。根据水资源的合理配置结果进行产业结构调整从而促进龙口市社会经济的持续发展。

参考文献

[1]　冯尚友. 水资源持续利用与管理导论[M]. 北京：科学出版社，2000.
[2]　王好芳. 基于量与质的多目标水资源配置模型[J]. 人民黄河，2004，26（6）：14-15.
[3]　陈晓宏，陈永勤，赖国友. 东江流域水资源优化配置研究[J]. 自然资源学报，2002，17（3）：366-372.

山东半岛地下水库建设与研究进展

刘青勇 张保祥 张 欣 王明森
山东省水利科学研究院，济南

1 前言

山东省是我国沿海省份之一，地处黄河下游，总面积约 15.67 万 km²。截至 2005 年年底，山东省总人口为 9 248 万人。山东省多年平均水资源总量为 308 亿 m³，人均占有量 333 m³，仅为全国人均占有量的 1/7。按现状供水能力分析，平均年缺水 80 亿 m³，枯水年缺水 150 亿 m³，据预测 2010 年缺水 230 亿 m³。由于水资源短缺，致使山东省平原区地下水严重超采，每年的超采量约 30 亿 m³，区域性地下水漏斗面积达 2 万 km²，有的地区发生了地面沉降，沿海地区出现了海、咸水入侵，面积达 1 000 多 km²。为了有效降低缺水造成的损失，山东省各地均采取了许多开源节流的措施。在山东半岛（鲁东低山丘陵水文地质区）地区兴建地下水库增蓄水资源就是众多措施之一。

山东半岛诸河流源短流急，主要降雨集中在汛期。为了有效利用暴雨洪水资源，同时阻止海水沿下游地下水含水层入侵；自 20 世纪 80 年代以来，山东半岛陆续兴建了龙口市八里沙河地下水库和黄水河地下水库、青岛市大沽河地下水库、莱州市王河地下水库、烟台市夹河地下水库等。2010 年前规划兴建的还有莱阳、沁水河、付疃河、平畅河、潍河、两城河地下水库，其地理位置见图 1。

2 典型地下水库建设与研究

2.1 龙口市八里沙河地下水库

为了防止海水入侵、缓解供水危机，"七五"期间山东省科委课题"八里沙河地下水库调节地下水（防止海水入侵）技术研究"，采用高压喷射灌浆技术在八里沙河建立了地下水库[1]。

八里沙河位于龙口市西部，流域面积 47.75 km²，为季节性河流。位于大陈家农场的河谷前的地下连续防渗墙建于 1987 年。库区出口宽 756 m，坝底平均埋深 8.5 m，最大埋深 24.2 m，帷幕板墙平均厚约 50 cm，汇流面积 14.7 km²。在坝前形成的地下水最大回水面积约 0.68 km²，最大储水空间为 42.97 万 m³，兴利库容达到 35.5 万 m³，多年平均拦蓄量

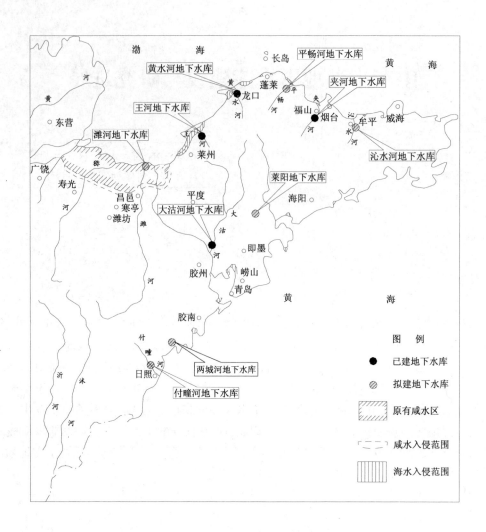

图 1　山东省海水入侵与地下水库分布

60 万～62 万 m^3，年复蓄指数为 1.8～2.0。

库区内有 62 个观测点、站，13 眼地下水开采井，坝前 10m 有 $150×606m^3$ 的方塘 1 个并配有机电提水设备。

2.2　龙口市黄水河地下水库

黄水河地下水库位于龙口市境内黄水河中下游平原区，库区北面濒临渤海，东西两侧为低山丘陵所环抱。龙口市黄水河地下水库于 1995 年建成，由以下工程组成[2]：

（1）回灌补源工程：通过在河床建设渗沟、渗渠、渗井等措施，与地下含水层沟通，增加地下水入渗量。模拟黄水河汛期洪水流经机渗井（有机钻施工的水泥管井）和人工组合渗井野外试验结果表明，单机渗井入渗量 $404m^3/d$，人工渗井 $76m^3/d$。为了强化地表径流向地下转化的能力，在中下游河道兴建了渗井 2218 眼、机钻渗井 300 眼、集水渗盆 773 个、集水渗沟 448 条，形成了渗井与机钻渗井相结合、集水渗盆与集水渗沟相结合、渗井与渗沟相结合的复合促渗系统。

（2）地表拦蓄工程：在黄水河主干河道中下游修建了 4 座大型水力自控式钢筋混凝土翻板拦河闸，对地表径流进行梯级拦截，与补源渗井工程相配合，组成地表、地下复合促渗补源工程。四座拦河闸一次性蓄水能力为 279.5 万 m^3，年复蓄利用水量 796 万 m^3。

（3）地下帷幕坝工程：在黄水河下游距海边 1.2 km 修建一座地下挡水坝，截面长度 5 996 m，平均坝高 26.7 m，最大坝高 40.1 m，最大坝深 43.4 m，截面总面积 16 万 m^2。地下坝具有拦截入海地下潜流和阻挡海水入侵地下水的双重作用。坝前形成的地下水库总库容 5 359 万 m^3，最大调节库容 3 929 万 m^3。

（4）提水工程：结合引渗工程，用集水廊道、辐射井、集水井等通过水泵提取地下水，送入供水管线。

（5）供水工程：在黄水河地下水库范围内，除了分散取水外，已经建成了四个集中供水水源地。

（6）排污工程：黄水河下游地区，供水工业比较发达，特别是乡镇企业较多，带来的工业污染也比较严重。为此，龙口市沿着黄水河的右岸铺设管道，对库区内工业排放的污废水采用管道式集中收集处理，达标排放。

（7）管理监测工程：地下坝两侧有 6 组 12 眼观测孔，库区范围内有 55 个地下水观测点。这些点除观测水位外，每两月进行一次区域性取水样，进行水质分析，重点地段进行水质加密监测。

2.3 青岛市大沽河地下水库

大沽河是山东半岛主要河流之一。全长 179 km。大沽河水源地 1986—1989 年开采实践；为了阻止库区下游的咸水入侵，保护库区水质，在地下水库的南端设置地下截渗墙，长 3.4 km，于 1998 年建成。

2.3.1 库区范围

大沽河地下水库位于该河中、下游河谷平原上，南北长 51 km，东西平均宽 5~8 km，总面积 421.7 km^2，以河谷平原的第四系冲积——冲洪积砂、沙砾石层为蓄水空间。

2.3.2 地下水库库容

大沽河地下水库的总库容是储水层顶、底板之间的存储空间，分区计算总库容为 3.84 亿 m^3，储水层最大厚度 15 m，平均厚度 5.19 m。

2.3.3 地下截渗墙

为了阻止库区下游的咸水入侵，保护库区水质，在地下水库的南端设置地下截渗墙，采用塑膜黏土防渗技术，长 3.4 km，最大深度 20 m。防渗墙右侧现为大沽河河床，建一大麻湾防潮橡胶坝，按抵御 10 年一遇大潮水位及下游农灌及压盐蓄水要求，坝高设计为 4 m，坝长 500 m，基础与截渗墙连成一体。

2.3.4 拦河闸及地下水回灌补源系统

在大沽河上共建设 7 座橡胶坝，坝高 2 m。回灌系统分为三个区。回灌系统包括干渠、支渠、斗渠和沉砂池和渗井。渗井设置在斗渠上，每平方公里 18~25 眼布置。渠底为砂层的斗渠不再设置渗井，渠道本身作为渗渠使用。研究表明有人工回灌措施时，大沽河地下水库的水量调节作用明显提高，年平均相差 2 600 万 m^3。

2.4 莱州市王河地下水库

王河位于莱州市东北部，在莱州市境内河长 50 km，总面积 326.8 km²。自 1977 年以来由于连续干旱，导致严重的海水入侵，1996 年海水入侵面积达 250 km²，占全市面积的 13.7%，占全市滨海面积的 44.1%，占山东省海水入侵面积的 40%。王河地下水库位于该市西北 15 km 处的王河下游，库区总面积 68.49 km²，地下水库总库容 5 693 万 m³，最大调节库容 3 273 万 m³。王河地下水库工程主要包括地下水库挡水坝工程、人工回灌补源工程、地下水开采工程、地下水监测工程等。

2.5 烟台市夹河地下水库

烟台市夹河是山东半岛主要河流之一，分为东西两支，流域面积 1 224 km²，中游建有门楼水库，设计库容 1.26 亿 m³。2000 年实施地下水截渗墙工程，工程总长度 4 030 m，形成地下水库容 2.05 亿 m³，调节库容 6 000 万 m³。其余工程与上述工程类似。

山东半岛主要地下水库技术指标见表 1。

表 1 山东半岛典型地下水库主要指标

河 流	流域面积/km²	库区面积/km²	含水层厚度/m	地下坝长度/m	地表拦水工程	库容/(万 m³)
龙口八里沙河	47.75	14.7	4~6	756	拦河闸	42.97
龙口黄水河	1 034	51.82	2~5	5 996	拦河闸	5 359
青岛大沽河	4 631	421.7	5~19	2 600	橡胶坝	38 400
莱州王河	326.8	68.49	6~36	13 500	拦河闸	5 326
烟台夹河	2 296	65	5~30	3 890	拦河闸	20 500

3 建设地下水库的潜力分析

山东半岛具有建设地下水库的优越自然条件，同时这一地区已初步形成了一套切实可行的建设地下水库的技术措施，取得了建设地下水库与回灌补源的综合效益。

（1）具有良好的水源条件。

山东半岛多年平均降水量 756 mm（1956—1998 年），扣除地表水利用量、入渗补给地下水量和植物截流等，每年约有 42.15 亿 m³ 地表径流直流入大海。因此，建设山东半岛地下水库的主要水源应是拦截直流入海的地表径流，在保证生态用水的前提下，力求最大限度地将地表径流转化为地下水库水源。

山谷河流下游冲积层和山前冲积平原地下水除接受大气降水、河流补给及地表水库调蓄外，还接受裂隙、基岩地下水的侧向径流补给。此外，山前冲积平原部分地区还有客水灌溉回归补给水源等。

（2）具有较好水质的补给水源。

山东半岛河流除东猪龙河的唐山段、白浪河的潍坊段、潍河的九台段、李村河的阎家山段、东五龙河的团旺段等污染较严重外，其他河流及河段作为地下水库的补给水源水质还是比较好的。

（3）巨大的地下储水空间。

山东半岛众多河流形成的山谷河流冲积层和山前冲洪积平原，分布面积广、数量多，含水层厚度较大、颗粒较粗，并且有较理想的隔水底板，地下水储存条件优越，形成巨大的地下储水空间，如青岛大沽河中下游冲积层形成的水源地砂层分布面积 293.24 km^2，砂层平均厚度 6.16 m，总体积 18.06 亿 m^3，其中可储水体积为 3.09 亿 m^3，相当库容 3 亿 m^3 大型地表水库的总库容量。

（4）良好的入渗条件。

山东半岛河谷冲积层和山前冲洪积平原受沉积条件的影响，除部分地段外，一般含水层以上的包气带较薄，沉积颗粒较粗，结构较松散，渗透能力较强，有利于降水和地表水的入渗补给地下水。特别是河谷冲积含水层分布区上覆盖层多为黏质砂土，而且沿河床多存在河床砂层与其下部砂层上下连通的"天窗"，局部地段砂层裸露，对地表水的渗透十分有利。

4 主要技术措施

按照地下水库所处位置的水文地质条件和工程设施等，将山东半岛地下水库可分为有坝地下水库和无坝地下水库两种类型。

4.1 有坝地下水库

有坝地下水库主要由拦截径流系统和引渗系统组成。拦截径流系统包括河床拦水坝和地下截渗坝，前者既可拦截河川径流，回灌地下水，又可防止海水沿河道上溯；后者既能拦截原来流向下游径流排泄的地下水，扩大地下水库的调蓄能力，又能有效地阻挡海（咸）水的地下入侵，均具"双阻水"作用。而引渗系统包括渗井、渗沟、渗渠等。

拦水坝一般应建在河床的较稳定段，要求坝址上游的河槽比较顺直、规则，上游河床、河滩开阔平缓，水流相对较平稳，坝下游河床宽阔不致产生回流，同时为使工程量小、投资省，坝址应选在河床较窄处。

4.2 无坝地下水库

当含水层水力坡降、透水性能均未达到有坝地下水库建设条件，或者虽然含水层厚度较大、分布范围较广却没有条件建立截潜流地下坝时，可直接建立无坝地下水库。无坝地下水库利用地下水水平径流缓慢的特点，用地表水回灌抬高地下水位，可以形成水丘，有利于提取。这种不建地下截渗坝的方式，同样可以形成地下调蓄水库。山东半岛西部山前冲洪积平原地形相对平坦，含水层厚且分布比较广泛，地下水径流条件差而水平方向径流又不占主要地位，因此这些地段适于建设无坝地下水库。

综上所述，山东半岛建设地下水库及利用汛期多余洪水调蓄地下水，不论是自然条件还是技术措施都是可行的，其潜力也是巨大的。区内投入运行的几个地下水库均发挥了不同程度的综合效益。

5 研究进展

5.1 拦蓄调节地下水技术研究

由山东省水利科学研究院承担完成的山东省科技厅课题"地下水库拦蓄调节地下水（防止海水入侵）技术研究"，是以龙口市八里沙河和黄水河地下水库建设为研究对象，主要研究内容如下[1]：

（1）地下建库规划设计计算参数。
（2）高喷建坝设计板墙厚度及孔距计算方法。
（3）地下水库调节计算方法。
（4）地下水库回灌补源工程措施。
（5）地下坝无损检测技术。
（6）地下水库经济效益。

该成果集地下水开发与水环境保护于一体，为滨海平原缺水地区供水开源、防止海水入侵修建地下水库提供了设计依据和建库方法。

5.2 地下水回灌补源模式研究与示范

由山东省水利科学研究院承担完成的山东省科技厅课题"地下水回灌补源模式研究与示范"，在以下方面进行了研究：

5.2.1 拦蓄补源及地下水库管理技术

（1）莱州市王河流域防止海水入侵的拦蓄补源工程技术。
（2）莱州王河地下水库与回灌补源工程开发利用。

5.2.2 地下水库系统工程与运行管理技术

（1）黄水河地下水库数值模拟研究。
（2）基于GIS的黄水河地下水库实时调度管理系统。
（3）引黄补源区无坝地下水库建设与管理。
（4）广饶地区地下水人工调蓄防治咸水入侵技术。
（5）邹平引黄补源地下水调蓄技术研究。

该成果在总结前人研究的基础上，建立了不同类型区和不同形式的地表水、地下水联合调蓄与地下水补源综合技术体系，取得了显著的环境、社会、经济效益。该项目总体思路框架见图2。

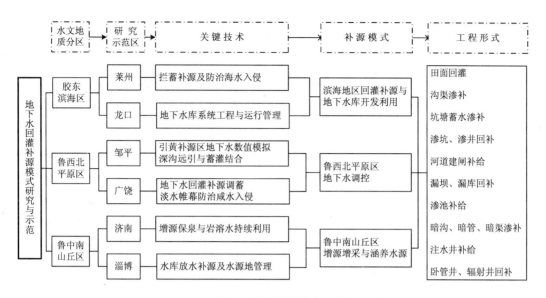

图 2　总体思路框架

6　结语

山东半岛水资源短缺，为增加水资源可利用量，加强地表水和地下水的联合调度，改善区域地下水环境，建立多处地下水库，并列专题在规划设计和管理方面进行深入的研究，取得了重要研究成果。研究表明，山东半岛建设地下水库及利用汛期多余洪水调蓄地下水，不论是自然条件还是技术措施都是可行的。区内投入运行的几个地下水库均发挥了不同程度的综合效益[3]。

由于地下水库建设还处于起步阶段，技术要求尚需规范。从建成的地下水库监测发现地下水环境污染问题已经显现，如硝酸盐积聚迅速，如何改进地下水设计和防止地下水污染，是亟待解决的课题。此外，地下水库理论体系的构建有待进行长期的观察和系统的专门研究。

参考文献

[1] 李道真,刘培民. 滨海平原地下建库供水开源防止海水入侵技术研究[J]. 水利水电技术，1997，28（8）：184-197.

[2] 刘青勇,马承新,张保祥,等. 地下水回灌补源模式研究与示范[J]. 水利水电技术，2004，35（2）：57-59.

[3] 徐军祥,康风新. 山东省地下水资源可持续利用研究[M]. 北京：海洋出版社，2001.

海水入侵防治研究与实践进展

李福林[1]　赵德三[2]　陈学群[1]
1. 山东省水利科学研究院，济南
2. 山东省农业委员会，济南

海水入侵是海岸地区的普遍问题[1]，国际上海水入侵的研究经历了静力学、渗流动力学和渗流—弥散动力学等研究阶段[2]，学术研究的领域也从原来单纯地对咸、淡水界面模型的模拟，发展到包括海水入侵的基本理论、数值模型、水文地球化学和环境同位素、研究和调查方法、防治和减缓对策、生态影响、全球气候和海面变化影响等多个方面[3,4]。

中国的海水入侵的研究领域主要集中在咸、淡水界面运移模型（范家爵，1988；吕贤弼，1991；薛禹群等，1991；艾康洪，1994；李国敏，1995；袁益让等，1996；陈鸿汉，2000等）、海水入侵的沉积环境（尹泽生，1992；庄振业，1996；孟广兰，1997；李道高，2000等）、水文地球化学和地球物理勘探（陈建生，1987；韩延树，1993；吴吉春，1996；邱汉学，1996；张永祥，1996；周训，1997；张保祥，1997；李福林，1999等）、灾害防治（赵德三，1991，1996等）以及大量的个例研究等方面。

海水入侵的研究主要是因实践的需要而推动发展起来的，因此，一开始便与防治实践结下了不解之缘。相对于基础理论研究和应用技术研究而言，目前中国海水入侵防治的实践，不单纯是一个水文地质学中的技术层面的问题，而是牵涉资源利用和环境保护的区域发展的重要课题，已早早从本本走到了活生生的现实，中国海水入侵防治理论的发展明显落后于实践的发展。因此，有必要对中国海水入侵防治的理论与实践作进一步的总结，以利于完善和提高。

1 中国海水入侵防治的发展阶段

根据海水入侵防治的理论发展以及实践深度，可把中国海水入侵的防治分为三个历史阶段。

1.1 调查和初步研究阶段（20世纪70年代后期—80年代中期）

最早发现海水入侵现象的是沿海地区使用机井灌溉的农民。20世纪70年代中期，为开发地下水资源，沿海和内陆一样，兴起了打井高潮。原来布局在天然咸、淡水界限附近的水井，只有当大潮或风暴发生时，井水才发生短暂的变咸，之后恢复。后来，随着农业灌溉面积扩大，水井密度的增加，开采量增加，某些水井水质长期变咸，浇地后造成庄稼枯萎乃至死亡。如山东省的龙口市和莱州市在 1975 年和 1976 年便有井水变咸无法恢复的调查报道。进入 80 年代，沿海乡镇企业异军突起，工矿企业发展迅猛，其水源地布局不合理，多建在咸、淡水界限附近，过量抽取地下水，加剧了海水入侵的发展，以致出现工农业争水打官司的局面。至 80 年代中期，海水入侵已成为十分普遍的一种现象，当地群众称为"海水倒灌"，调查和初步研究由当地水利部门和农业部门完成，调查材料上多称为"海水浸染"。

这一时期，对海水入侵的防治是一种被动式的防御。井水变咸了，老百姓弃井，改向淡水一侧继续打井使用；浅层水变咸，改用深层水。企业行为也是如此，如山东省寿光市羊口盐场的自来水供水水源地初期建在王高镇三号县路的北侧，后因水质变咸，一再向南侧的淡水区迁移，由于地下水漏斗的牵引，咸水入侵速度达到 350m/a。

因此，该时期的海水入侵防治是一种无序的被动式的防御。

1.2 研究、试验和探索阶段（20 世纪 80 年代中期—90 年代中期）

该阶段由于前期无序的开发以及被动的防御，造成海水入侵的加速和大面积发展，山东省海水入侵区由局部地段不断扩大成片，可划分为潍北平原咸水入侵区、蓬、黄、掖滨海平原海水入侵区以及烟台、威海、青岛、日照等地市的众多河口地段海水入侵区。该时段也是我国"七五"和"八五"建设同步时期，特别是 80 年代末期，莱州湾沿岸一些市县粮食减产，工厂被迫减产、停产，造成了极大的危害。海水入侵灾害引起了省级政府和中央政府的高度重视。

国家安排了"七五"攻关第 57 项（75～57）和"八五"攻关项目"海水入侵防治试验研究"（85～806），部分高校和研究单位还从事了基金项目及其他项目的研究工作，研究区域从山东省扩大至辽宁、河北、福建和广西等全国沿海各地，从沿海扩展到海中的岛屿。研究领域从海水入侵的起源、成因、类型和机理，到海水入侵的渗流、弥散、扩散的路径和基本规律等各个方面，取得了大批研究成果，尤其是数值模拟的研究达到了较高水平和影响。该时期与海水入侵防治密切相关的理论认识主要有：

（1）现代海水入侵和咸（卤）水入侵的类型不同，防治措施也应不同。
（2）海水入侵与沉积环境相关，应突出重点，区别防治。
（3）海水入侵的发展存在阶段性变化规律，应采取分阶段防治措施。
（4）海水入侵防治要多项措施相结合，综合防治。

以赵德三、尹泽生等为代表的攻关项目组，提出综合防止海水入侵的措施，并付诸实际行动。具体包括兴建防潮堤、拦蓄补源、淡水帷幕、地下水回灌、引调客水、节水以及适应性作物栽培的农业开发等。其成果集中反映在几部著作中[5, 6]。

此外，一些学者还根据国外的经验，介绍了抽注水帷幕等防治措施，但该时期对于海水入侵的防治，处在研究、试验和探索阶段，上述防治措施仅限于技术层面，未上升到理论层次。

值得一提的是，山东省水科院在考察日本海水入侵的防治经验后，进行了地下坝防止海水入侵的工程试验研究，在龙口市八里沙河建成中国第一座防止海水入侵的试验坝，探索了一条利用地下防渗墙防止海水入侵的成功路子，该项成果获 1996 年度国家科技进步三等奖。

1.3 防治工程全面实施阶段（20世纪90年代中期至今）

从"九五"开始，国家对资源环境领域的科技投入重点方向逐渐转向中国西部，仅有为数不多的几个国家自然科学基金项目支持相关研究，该时期海水入侵防治试验的研究工作也基本告一段落，但地方政府对防治工程投资的支持力度增大。除了常规的拦蓄补源、地下水回灌防治工程，结合水利灌溉工程每年汛期实施外，这一时期还出现了一些新型的海水防治工程。

1.3.1 填海造陆式的防潮和蓄淡工程

山东省莱州市朱家村1994年底至1998年，为防止海水沿地表和河口的入侵，投资1166万元填湾造陆，分别建成长800m和3000m的两道拦海大坝，与普通防潮堤工程直接建在岸边不同，该村是将海岸线平直外推1500m，形成荷兰式的填海造陆防潮工程。此后又相继投资在围坝内开挖临海浅塘水渠、兴建淡水水库、栽植果树林和芦苇，营建人工湿地。该村经过几年的治理，不仅成功防治了海水入侵，还达到了较好的社会、环境效益。

1.3.2 河口地下坝工程

是指在滨海平原河口地区，主要采取高压喷射灌浆、静压灌浆等方法，构筑地下防渗墙，形成地下拦水坝，拦蓄地下潜流，以达到提高地下水位，防止海水入侵的目的。地下坝与其上游段的拦蓄工程常常结合形成所谓的地下水库，能够有效阻断海水入侵兼顾蓄水、供水。地下坝一般选择在河流入海处的咸、淡水界线附近，构筑的坝基下切不透水岩层，灌注材料选择防水性很强的水泥，大坝厚度为30～50cm，坝长可达数千米。目前全国已建有不同规模的地下水库7座，其中山东省建有八里沙河、黄水河、白沙河、大沽河、外夹河和王河等6座地下水库，辽宁省建有龙河地下水库1座。地下坝和地下水库工程，具有提高地下水位，防止海水入侵、扩大淡水供应、不占耕地、无垮坝风险、投资小等优点。

1.3.3 潮间带抽咸养殖工程

在潮间带抽取地下咸水养鱼是中国水产科学院黄海水产研究所的雷霁霖先生首先倡导的[8]。1992年他把原产于欧洲大西洋东北部沿海的"大菱鲆"引进中国并培育成功。1998年首先在山东省莱州市的朱旺村创办起"温室大棚＋深井海水"的养殖模式，此后，由于效益显著，逐渐推广到整个莱州市和山东省烟台的其他县市及日照市。现在，大菱鲆的养殖已从环渤海地区辐射到江苏、上海、浙江、福建、广东等沿海省市，形成了年产15亿元的新兴产业。在潮间带抽取地下咸水养殖，由于取水量巨大，降低了咸水水头，客观上起到了阻止海水沿地下含水层入侵的目的。以莱州市为例，现有养殖大棚1000多个，每个大棚日用水700～800m^3，全市每天抽取地下咸水量高达70万～80万m^3，已超出地下淡水的开采量。这样，必然打破原先的入侵动态，据初步观测，莱州朱旺剖面的海水入侵区有变淡的趋势。

2 海水入侵防治的理论与实践经验总结

2.1 现有海水入侵防治理论

从上述发展阶段的各项措施，我们总结出以下几种防治理论。

（1）入侵类型差别防治论。中国存在现代海水入侵和咸卤水入侵两种类型，防治应分别针对不同的入侵源采取不同的措施。该理论对于制定海水入侵的宏观战略规划时，具有较强的优势。

（2）入侵层次防治论。咸水赋存于多层的海相地层中，由于开采层的变化，存在顺层、多层和越层的入侵，针对这种情况，应当根据不同层位的入侵实际情况，对重点层位进行防治。该理论对于技术层面，控制海水入侵的越流入侵、串层入侵有较为有效的指导意义。

（3）灾害阶段防治论。蔡祖煌和庄振业分别提出咸（卤）水和海水入侵发展的阶段性变化规律[2, 7]，认为海水入侵从发生到发展，经历初始、加剧、减缓等三个阶段，该理论认识对于防治不同阶段的海水入侵具有指导意义。

（4）地下水限采论。基于地下水超采是引发海水入侵的主要原因，提出沿海地带要统一规划，限制地下水开采量。该理论对于制定省市级地下水开发规划，划定禁采区、限采区具有重要指导意义，已经在全国部分省市开始实施。

（5）人工补源论。主要包括拦蓄补源和地下水回灌，目的旨在抬高地下淡水水位。该理论指导的是一种常规的水利工程措施，只要合理利用，其成效十分显著，特别是对于入海流量巨大的河流，应当积极提倡。

（6）刚性工程论。主要指建造防潮堤和地下坝工程等刚性工程措施，通过直接切断淡水与咸水的联系来达到控制海水入侵的目的。

（7）客水论。该理论基于北方水资源短缺的现实和未来用水增长的需求，认为只有利用客水才是解决用水和防止海水入侵的根本措施，实际上属于水资源优化配置范畴。山东省沿海地区的成功经验和规划方案是，"限制开采地下水，扩大利用黄河水，积极引用客水"。

（8）抽咸补淡论。该理论最早由中国地质科学院水文地质环境地质研究所在华北地区进行综合治理旱涝碱咸的研究和试验中提出，目标是在地下咸水分布区，开采浅层苦咸水，腾出地下库容，利用天然降雨进行补给，以达到合理利用咸水资源，逐步降低土壤含盐量的目的。引申到海水入侵防治范畴，即是所谓的莱州市朱旺村和朱家村的模式，在咸水区大量抽取地下咸水，形成抽水帷幕，在淡水区兴建淡水水库，以淡压咸，从而达到防止海水入侵的目的。

2.2 实践经验及存在的问题

目前，山东省沿海已经初步形成海水入侵的综合防治体系（图1）。在具体实践过程中，海水入侵得到治理效果较好的是拦蓄补源、地下水回灌措施以及地下水库的建设，应用地区主要是河口地区和现代海水入侵区域。比如，山东省的寿光市的弥河、莱州市的王河均利用汛期大量雨洪，实施引水回灌，抬高地下水位；龙口、莱州、青岛、烟台等地通过地下坝截渗，阻隔咸水向淡水区入侵，均取得了较好的效果。治理区域往往会采取多种措施结合，以达到综合防治的目的。

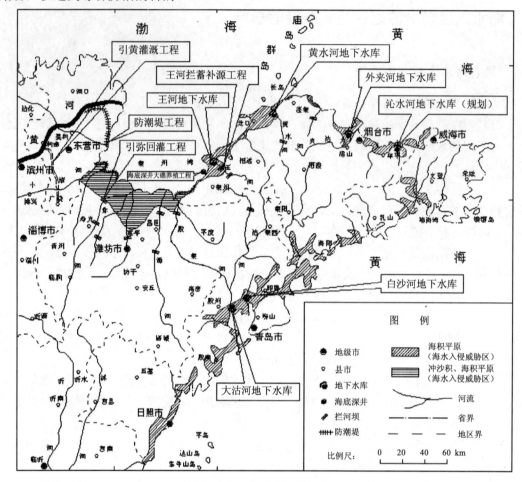

图1　山东省海水入侵区典型防治工程布局图

但是，目前在防治工作中还存在许多问题，主要表现在：

(1) 对承压含水层和多层含水层的入侵，尤其是咸（卤）水地区，第四系埋藏厚度较大，存在多个咸水层入侵问题，目前还缺乏有效的防治实践措施。

(2) 防治工程的规划和设计沿用老规范标准，缺乏创新和系统性。

(3) 措施比较简单，缺乏高技术含量。

(4) 现有的防治工程和措施，目标单一，较少考虑对生态环境的影响，同时对海水入侵的跟踪监测和治理工程的环境后效研究也很不够。

这一切，使得我们海水入侵防治工作还处在一个较低的层次。因此，如何在不违反安全性和海水入侵防治效果的基础上，拟出一个可以兼顾生态和环境的新型措施，将是未来我国进行相关海岸防护工程的新趋势。需要提倡新的防治理念。

3　海水入侵的生态防治论

在海水入侵防治上的两种极端论点，即任其入侵至不透水岩层的消极论，以及完全把地下咸水赶出陆地的积极论,均是不明智和不经济的做法。鉴于目前海水入侵防治理论上的不完善,防治理论以单一论居多,各有适用范围,未形成综合体系,以及较少顾及生态和环境的现状。我们借鉴国外海岸防护和河流生态恢复的先进理念[9, 10],提出生态防治理论的观点,以期对海水入侵防治理论的发展有所裨益。

所谓海水入侵的生态防治理论,是以防止海水入侵和海岸生态恢复为双重目标,综合运用各种工程的和非工程的措施,集成一个立体的防治体系,维持咸、淡水界面达到一种动态的平衡,把海水入侵控制在一个可容忍的程度。

为此,我们提出如下几点可供操作的具体措施。

3.1 构建生态型海岸防护建筑物

传统的海岸防护工程,不外乎兴建离岸堤、突堤和防潮堤等,将海岸线变成一道固若金汤的"海岸长城"。这种"水泥化"的海岸防护工程,不但需花费巨额的建造及维护费,使海岸景观变得极为单调,剥夺了民众亲海的空间,也破坏了珍贵的海岸生态环境。

从生态防治理论出发,要废弃直立式防潮堤的传统设计理念,代之以斜坡式,靠近水面部分因风浪作用力大可放置混凝土消波块,水面下则放置石块,这种护岸可减轻地基的承载力且造价低廉,重要的是这种抛石护岸具有鱼礁藻场效果。

在无碍航行安全下,可采取不破坏景观的离岸潜堤设计,或以适当宽度的海滩沙丘及种植沙丘植物替代消波块,以增加海岸的自然景观性,及维护海岸的栖息地,并能保护海岸的生态环境。

另一种新的做法,是二重堤或主副堤的设计,即在主堤前面另设置一复断面的倾斜堤,取代主堤前的大量消波块,复断面堤前面的消波块可减小数量或体积,不需完全阻挡波能。复断面堤后面的小段部,有适当的水深和足够的宽度可生长海藻、海草。而中间水域因波浪变小可成为很好的生物栖息环境。

3.2 地下水库设计和建设要考虑环境影响

目前国内修建的地下水库,均是建在不透水基岩上的"封闭式"或"贮存型",地下坝拦截了地下潜流,造成地下水在库内循环速度变缓,一旦上游污水流入,水质的恢复需要较长的时间。对此我们可以采取强排的方法处理,但要浪费很大的人力、物力。从生态防治的角度出发,最好设计不建在基岩上的"开敞式"、"半封闭式"或"径流调控型"的地下水库,以利于地下水的循环和物质交换,当然,要加强监测和用水管理,保证库内地下水位降低的幅度不至于引起咸水的入侵。这在咸水入侵的非敏感地段和危险性相对较差地段可以采用。另外,设计和建设时还要考虑地下水位升高会不会带来坝基的不稳定及由此产生的滑坡,地下水位接近地面能否带来对植被和作物的不利影响等。

3.3 合理规划布局,潮间带适度开采地下咸水

个别沿海地区为了单纯经济利益,形成一哄而上的抽咸养殖的局面,地下水位大幅度下降,水量出现明显不够,倡导咸水养殖的雷霁霖也认识到了这一点。大功率内燃机车开进海滩打井,大量土石料和人工垃圾随处弃置,造成近海新一轮污染,而且金灿灿的砂质海岸养殖大棚林立,防护林被沙坑包围。潮间带生物资源丰富,海滩是吸引游客的旅游场

所，人类活动的增强对海岸生态系统和栖息地的威胁，必然引起生物多样性的丧失及对自然环境的破坏。另一方面，滨海淡水含水层和近岸地下咸水层是一个统一的、陆海相互作用的水文地质单元，地下咸水的大幅度开采，在引起咸淡水界面后退的同时，咸水区由于抽水井密封不严、废井的"开天窗"等人为因素，接受上覆海水的补给量也会大大增强，如果大规模的海水涌入地下咸水层，能否引起新一轮的海水入侵尚属未知。因此，我们提倡咸水开采要在计算允许开采量基础上，因地制宜，统一规划布局，加强管理。

3.4 河道用水，要留有入海径流，保护河口海岸湿地和生态

河流生态恢复技术是一种新的治河理念，在满足人对开发利用水域需求的同时，还兼顾水体本身存在于一个健全生态系统之中的需求，同时，河流恢复工程建设还提倡公众对于水环境保护的积极参与，造成一种人与自然亲和的环境，能够体现水域天然的美学价值。

根据这一观点，在开发利用河流时，要明确河流与其上下游、左右岸的生物群落处于一个完整的生态系统中，进行统一的规划、设计。在满足用水要求的基础上，尽可能维持一定的入海径流，保护河口湿地，保护海岸植被群落，不仅能够有效缓解海水入侵，还可为当地野生的水生与陆生植物、鱼类及鸟类等动物的栖息繁衍提供方便条件，改善生态环境，更好协调人们在用水与生态系统建设的关系。

参考文献

[1] Custodio，E. Groundwater problems in coastal areas.UNESCO，Belgium，1987.

[2] 蔡祖煌，马凤山. 海水入侵的基本理论及其在海水入侵发展预测中的应用[J]. 中国地质灾害与防治学报，1996，7（3）：1-9.

[3] 李国敏，陈崇希. 海水入侵研究现状与展望[J]. 地学前缘，1996，3（1，2）：1-5.

[4] Alexander H-D. Cheng, Leonard F. Konikow and Driss Ouazar, Special Issue of Transport in Porous Media on 'Seawater Intrusion in Coastal Aquifers', *Transport in Porous Media*, 2001，43：1-2.

[5] 尹泽生. 莱州市滨海区域海水入侵研究[M]. 北京：海洋出版社，1992.

[6] 赵德三. 海水入侵灾害防治研究[M]. 济南：山东科技出版社，1996.

[7] 庄振业，刘东雁，杨鸣，等. 莱州湾沿岸平原海水入侵灾害的发展进程[J]. 青岛海洋大学学报，1999（1）：141-147.

[8] 雷霁霖，门强，王印庚，等. 大菱鲆"温室大棚+深井海水"工厂化养殖模式[J]. 海洋水产研究，2002，23（4）：1-7.

[9] Ronald M. Thom, Adaptive management of coastal ecosystem restoration projects, Ecological Engineering . 2000，15（3-4）：365-372.

[10] 董哲仁. 生态水工学的理论框架[J]. 水利学报，2003，1：1-6.

莱州湾海（咸）水入侵区宏观经济水资源多目标决策分析模型

何茂强[1]　王维平[1]　黄继文[2]　范明元[2]
1. 山东济南大学资源与环境学院，济南
2. 山东省水利科学研究院，济南

1　背景

莱州湾海（咸）水入侵区沿莱州湾海岸呈半环状，东西长 200 km，南北宽约 40 km，行政区划包括烟台市的龙口、招远、莱州，潍坊市的昌邑、寒亭、寿光，东营市的广饶及青岛市平度的新河、灰埠、官庄三个乡（镇），总面积 10 022 km²，总人口 462 万人，耕地面积 4.2×10^5 hm²，国内生产总值 228 亿元，粮食产量 3.14×10^6 t，总供水量 1.278×10^9 m³，其中地表水 2.725×10^8 m³，地下水 10.059×10^8 m³，工业用水 1.255×10^8 m³，农村用水 10.689×10^8 m³，城镇生活用水 0.83×10^8 m³。20 世纪 80 年代以来由于连续干旱，加之用水量增加，地下水长期过量开采，导致大面积海（咸）水入侵，入侵面积已发展到目前的 733.4 km²，对当地工农业生产、人民生活和生态环境带来了严重的危害。莱州湾海（咸）水入侵是水资源缺乏造成的，但引起该地区水资源危机深层次的原因是当地社会经济发展与人口的增加，这种发展和水资源的开发已超过了自然的承受力，因此，不能孤立地研究海水入侵问题。莱州湾海（咸）水入侵区宏观经济水资源多目标决策分析模型，用系统分析的方法，以水资源作主线，把区内宏观经济、海水入侵紧密结合在一起，将三者作为一个大系统研究。在这个系统内，三者之间相互影响、相互关联，形成一个十分复杂的系统。

2　模型范围及设计任务

项目研究区包括莱州湾海（咸）水入侵区的 8 个市（县、区），在规划时间上分 3 个水平年：1993 年、2000 年、2010 年。根据经济分析和需水分析的要求，将国民经济分为第一产业（农业）、第二产业（工业和建筑业）、第三产业（交通运输业、邮电业、服务业和其他）。考虑到农业是用水大户，对该区海水入侵影响很大，又将农业进一步划分为：种植业和非种植业（林、牧、副、渔四业），其中种植业又分为旱地作物与灌溉作物，而灌溉作物又分为冬小麦、早秋、晚秋、经济和蔬菜五种作物。

多目标决策分析模型的设计任务是[1]：

（1）提供一个决策者和专家与计算机的对话界面，以辅助决策者和专家进行决策。

（2）协调宏观经济与社会发展及海（咸）水入侵之间的矛盾与冲突，提出各水平年理想的、可行的、非劣的经济、水环境和社会发展目标，并使之与水资源开发相适应。

（3）模拟决策者和专家进行决策，提供在具有水资源、海（咸）水入侵及其他约束条件下经济发展的趋势及数据。

（4）提供不同目标权重（即不同发展模式）下的产业结构分析和各产业增加值的预测值。

（5）分析"引黄济烟"工程早上与晚上对该地区产生的影响，以及水资源的供需平衡和优化配置。

3 模型目标的选择及约束条件的确定

3.1 经过大量调查研究，确定采用三个目标[2]

（1）经济目标 GDP：三个水平年平均 GDP 最高。GDP 是一项全面反映经济活动的指标。国内外在进行比较分析时，则常采用人均 GDP 这一指标，并且为主要指标。

（2）社会目标 FOOD：三个水平年人均粮食占有量最高。粮食占有量不仅是居民生活水平的标志，同时也是一个不容忽视的社会指标。由于人多地少，加之人口增加，非农业生产占地增加及海水入侵造成耕地减少，对粮食生产影响很大。同时，粮食的生产需要消耗大量的水，它与水资源配置和海水入侵密切相关。

（3）海水入侵目标 SIA：三个水平年平均海（咸）水入侵面积最小。海水入侵是该区面临的主要自然灾害，因此决策者必须在经济发展与海水入侵加重两方面做出切实可行的权衡与决策。

以上三个目标可以大体上反映出区域经济水环境与社会协调发展的基本情况。这些目标显然是相互竞争的，增加一个目标必然以牺牲其他目标为代价。多目标决策分析模型的目标及其主要变量的关系如图 1 所示。

3.2 多目标决策分析模型的约束条件[3-5]

3.2.1 宏观经济约束

（1）GDP 与各部门增加值以及各部门增加值与产值的关系。

（2）投入—产出关系。

（3）组成最终需求的居民消费和社会消费，固定资产与流动资产的比例关系，以及出口与进口差额的上、下限及其在不同阶段的变化规律。

（4）总投资与固定资产投资、流动资金及水投资之间的关系。

（5）固定资产投资与固定资产原值的关系。

（6）固定资产原值与产值之间的关系。

（7）农业产值的形成。

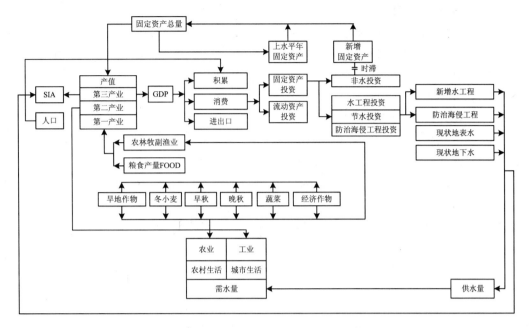

图 1　多目标决策分析模型的目标及其主要变量关系

3.2.2　农业约束

(1) 农业结构：农业种植业与非种植业产值的关系。
(2) 灌溉面积：旱地面积和总耕地面积的关系及变化。
(3) 作物的种植结构及其变化。
(4) 粮食产量与播种面积、粮食单产的关系。
(5) 人均粮食占有量。
(6) 种植业产值的形成，林、牧、副、渔业产值的形成。

3.2.3　水量平衡约束

(1) 工业需水与产值的关系。
(2) 农业需水与播种面积的关系。
(3) 城镇及农村生活需水和人口的关系。
(4) 地下水可供水量的约束。
(5) 地表水可供水量的约束。
(6) 新增工程供水能力约束。
(7) 需水总量与供水总量的关系。

3.2.4　水投资平衡关系

(1) 农业节水量的形成。
(2) 工业节水量与工业产值的关系。
(3) 工业节水投资与工业节水量的关系。
(4) 新增工程的投资与工程的关系。
(5) 水投资总值的形成。
(6) 水投资总值与宏观经济中可提供的水投资之间的关系。

3.2.5 水环境平衡约束

（1）需水量与地下水开采量的关系。

（2）海水入侵速度与地下水开采量的关系。

（3）海水入侵速度的上、下限。

（4）海水入侵面积与耕地面积的关系。

3.2.6 水利工程关系约束

（1）所有的工程都用0~1变量表示，一个工程在某个水平年存在，其相应的0~1变量值取1，否则取0，工程之间的关系则是这些0~1变量的关系。

（2）同一个工程上、下水平年的关系。如果同一个工程在某一个水平年存在，则在下一个水平年继续存在。

（3）不同工程之间的关系，主要包括工程上马的先后次序及其他关系。

3.2.7 切比雪夫方法约束

（1）目标的归一化，矢量纲化。

（2）切比雪夫距离的形成。

（3）切比雪夫方法目标函数的形成。

图2揭示了目标与主要约束和数据之间的主要关系。莱州湾海（咸）水入侵区宏观经济水资源多目标决策分析模型，涉及8个区，3个水平年，整个模型由2550个方程2000多个变量组成。

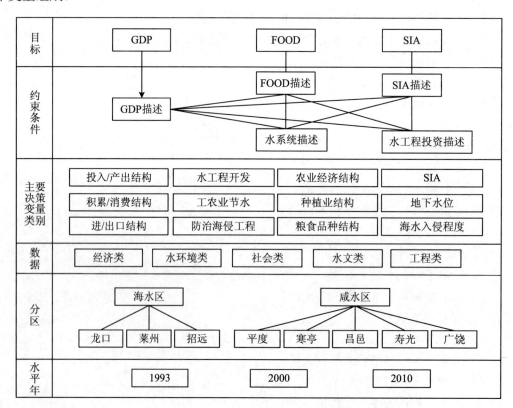

图2　多目标决策分析模型结构示意图

4 多目标决策分析模型的切比雪夫解法

切比雪夫（Tchebycheff）方法是美国佐治亚大学经济管理学院教授 Steuer 于 1983 年提出的，是一种通过与决策者的交互而逐渐缩小决策空间、最终达到满意解的方法。它的求解过程是通过随机分配理论，抽取若干组编好结构的模型解，并将隐去权重结构的这几组解提供给决策者挑选，决策者在权衡目标间的得失关系以后，从中挑选出决策者"第一次满意解"，Tchebycheff 过程以这个解为核心，进行离散和抽样，又给出若干组新的权重结构下的模型解，供决策者挑选。这种交互过程是反复进行，直至求出决策者的满意稳定的解，或者再也不可能产生其他的新解为止。每经过一次交互，都将诱导决策者的判断意愿和偏好于模型求解过程中，并将搜索满意解的空间逐步缩小，直至求出决策者的满意稳定解为止。这一搜索过程能使决策者有充分的机会参与决策过程，并有效、完整地表达自己的思想、偏好和意愿。

5 模型的优化结果

鉴于该地区缺水和海水入侵严重，实现跨流域调水势在必行。下面着重对两个方案，即"引黄济烟"工程 2000 年生效方案（简记为 YH2000 方案）和"引黄济烟"工程 2010 年生效方案（简记为 YH2010 方案）进行了多目标分析，优化结果：YH2000 方案多目标分析优化结果的 GDP 值及增长率见表 1，YH2000 方案多目标优化结果的粮食产量和人均粮食占有量见表 2，YH2000 方案多目标优化结果各水平年海（咸）水入侵面积见表 3，YH2000 方案多目标优化结果各地区需水量分配见表 4，YH2000 方案多目标优化结果的各地区供水量见表 5。

表 1 YH2000 方案多目标分析优化结果的 GDP 值及增长率

市（县、区）	国内生产总值 GDP/（$\times 10^6$ 元）			国内生产总值增长率/%		
	1993	2000	2010	1993—2000	2000—2010	平均
龙口	4 266	9 136	24 760	11.50	10.48	10.90
莱州	5 023	10 753	29 088	11.49	10.46	10.88
招远	3 733	7 986	21 566	11.48	10.44	10.87
平度	192	381	782	10.30	7.46	8.62
昌邑	2 963	5 949	14 766	10.47	9.52	9.91
寒亭	1 727	3 484	8 671	10.55	9.55	9.96
寿光	4 391	8 723	21 226	10.30	9.30	9.71
广饶	1 052	2 059	4 865	10.06	8.98	9.43
合计	23 347	48 471	125 724	11.00	10.00	10.41

表2 YH2000方案多目标优化结果的粮食产量和人均粮食占有量

市（县、区）	粮食产量/（×10³t）			人均粮食占有量/（kg/人）		
	1993	2000	2010	1993	2000	2010
龙口	360.57	356.76	345.64	599	531	478
莱州	505.88	466.98	445.02	576	473	419
招远	334.18	297.94	324.54	284	471	477
平度	69.17	53.76	53.50	948	656	608
昌邑	511.46	473.98	459.04	763	629	567
寒亭	277.52	285.77	268.23	788	728	627
寿光	733.07	699.82	673.27	724	626	560
广饶	357.64	366.48	375.07	772	724	689
合计	3149.49	3001.49	2944.30	681	583	532

表3 YH2000方案多目标优化结果各水平年海（咸）水入侵面积

市（县、区）	海（咸）水入侵面积/（×10³hm²）			海（咸）水入侵速度/（m/a）	
	1993	2000	2010	2000	2010
龙口	9.08	9.74	10.70	24	24
莱州	23.42	25.89	30.20	201	247
招远	1.56	3.10	5.87	37	46
平度	5.6	5.68	5.79	7	6
昌邑	8	9.13	9.21	43	2
寒亭	8.51	9.12	9.54	44	21
寿光	11.04	19.84	31.46	359	332
广饶	2.04	2.41	2.93	14	14
合计	69.25	84.91	105.70	84	78

表4 YH2000方案多目标优化结果各地区需水量分配表　　单位：×10⁶m³

部门	水平年	龙口	莱州	招远	平度	昌邑	寒亭	寿光	广饶	合计
工业	2000	41.48	54.45	50.45	1.58	18.21	28.15	40.51	35.57	270.4
	2010	83.21	110.38	125	4.51	42.52	52.41	71.05	64.37	553.45
种植业	2000	144.92	206.65	163.27	26.2	240.36	155.26	283.39	276.13	1496.18
	2010	150.36	212.13	175.21	27.85	251.39	163.79	292.45	287.05	1560.23
非种植业	2000	5.29	4.47	2.53	0.44	3.48	4.74	5.28	4.75	30.98
	2010	12.56	11.22	2.21	1.1	8.78	12	13.58	12.07	73.52
城市	2000	4.52	3.8	2.41	0.16	2.68	3.21	3.32	1.61	21.71
	2010	7.17	7.12	4.56	0.44	5.42	4.72	7	3.28	39.71
农村	2000	9.17	14.48	9.3	1.28	11.25	5.08	16.84	7.58	74.98
	2010	10.23	16.16	10.38	1.43	12.55	5.67	18.79	8.45	83.66
合计	2000	205.38	283.85	227.96	29.66	275.98	196.44	349.34	325.64	1894.25
	2010	263.53	357.01	317.36	35.33	320.66	238.59	402.87	375.22	2310.57

表5 YH2000方案多目标优化结果的各地区供水量　　　　　单位：$\times 10^6$ m³

水源	水平年	龙口	莱州	招远	平度	昌邑	寒亭	寿光	广饶	合计
地表水	2000	70.4	77.8	63.72	2.53	163.04	103.86	46.27	226.06	753.68
	2010	70.95	78.02	64.66	2.78	189.41	112.77	55.02	237.82	811.43
地下水	2000	93.91	134.69	102.76	14.97	102.72	73.08	256.4	92.92	871.45
	2010	96.53	135.19	109.99	16.97	107.72	78.08	265.4	94.66	904.54
污水	2000	7.05	9.69	9.11	0.28	2.94	5.18	7.02	6.65	47.92
	2010	19.57	27.4	27.73	0.79	8.24	13	10.61	18.53	125.87
地下水超采	2000	21.28	1.67	37.37	3.88	7.18	14.32	39.65	0	125.35
	2010	25.28	44.4	99.93	7	0	0	71.85	24.21	272.67
工程	2000	12.73	60	15	8	0.1	0	0	0	95.83
	2010	51.2	72	15.05	7.79	15.29	34.74	0		196.07
合计	2000	205.37	283.85	227.96	29.66	275.98	196.44	349.34	325.63	1894.23
	2010	263.53	357.01	317.36	35.33	320.66	238.59	402.88	375.22	2310.58

6 结论和建议

（1）该区是缺水地区，水资源和海水入侵对经济发展起着制约作用。1993—2010年全区GDP平均年增长率达到11%是有可能的，但由于水资源、海水入侵及其他方面的影响，不可能超过14%。全区第一产业和第二产业占GDP的比例逐渐下降，第三产业占GDP的比例稳步上升，并且升幅较大。从增长率来看，第一产业增长率较慢，第三产业发展速度最快，第二产业发展居中，其中第一产业中非种植业发展最快。从全区GDP增长率的时段分布来看，后期比前期有所降低，主要原因是前期经济总量处于较低水平，水资源对经济的制约还不是很大，但当经济的增长近似于指数时，而新的水源工程的增加有限，节水也有限，因而水的供需缺口较大，这样，水资源的平衡主要是靠降低经济发展速度来维持。

（2）莱州湾海（咸）水入侵区主要以农业生产为主，农业用水占总用水量的80%以上。粮食生产占有重要地位，而粮食生产用水量大，约占农业用水的80%，因此粮食生产对区内的GDP发展速度和海（咸）水入侵速度影响最大。模型运行优化结果表明，追求粮食产量最大，经济目标降到最低点，海（咸）水入侵面积达到最大值。受非农业占地和海（咸）水入侵的影响，耕地面积将减少，灌溉面积和农作物复种指数略有增加，经济作物和蔬菜种植面积逐渐增加，粮食总产量呈下降趋势，将由1993年的3.15×10^6t下降到2000年的3.0×10^6t和2010年的2.94×10^6t；人均粮食占有量由1993年681kg，到2000年和2010年分别下降到583kg和532kg。从水资源供需平衡、维持经济持续发展和进一步防治海（咸）水入侵速度加快的角度考虑，特枯年和连续枯水年应限制粮食生产，适当调入部分粮食。建议区内粮食生产保持在满足人均400kg左右，粮食无须调出。

（3）"引黄济烟"工程上马易早不易迟。2000年"引黄济烟"工程上马对该区经济发展和缓解海（咸）水入侵有显著的作用。根据模型运行结果，"引黄济烟"工程若迟上10年，预测区内的GDP 2000年和2010年将分别减少24亿元和75亿元，并且海（咸）水入侵面积将进一步扩大。

（4）模型输出结果表明，在一定经济发展速度和粮食产量条件下，海（咸）水入侵面积由现在的 690 km², 到 2000 年和 2010 年将分别增加到 850 km² 和 1 060 km²，但海（咸）水入侵的速度是逐渐降低的。这主要是由于一些地表拦蓄补源工程和"引黄济烟"工程的上马，在一定程度上缓解了地下水的开采量。但莱州湾地区由于地下水超采过量，加上经济还需要继续增长，所以海（咸）水入侵仍将继续发展。

参考文献

[1] 王维平，等. 烟台市宏观经济动态投入产出模型[J]. 中国人口·资源与环境，1994（4）.

[2] 王维平，等. 宏观经济水资源投入产出分析模型[J]. 水利经济，1995（2）.

[3] 翁文斌，等. 宏观经济水资源规划多目标决策分析方法研究及应用[J]. 水利学报，1995（1）.

[4] Noth，R.M.，Cool Programming Applications in Water Resources Project Analyses，Multi-objectives Analysis in Water Resources（Edited by Haimes，Y.Y. and D.J.Allee，American Society of Civil Engineers），1982，Vol.3.

[5] Haimes，Y.Y. and W.A.Hall，Multi-Objectives in Water Resources Systems Analysis：The Surrogate Worth Trade off Method，Water Resources Research，1974，Vol.10，pp.615-624.

流域生态需水
——洪水的重要性

贠汝安
山东大学土建与水利学院，济南

1 洪水是什么

我们通常将洪水定义为"将一般干地淹没的情况"。在许多河流系统中，有明确界限的河道和邻近平地（漫滩）之间是有严格区别的。当河流中的水流超过了河道的容量，并且泛出漫滩时，洪水通常就会发生。然而，在其他一些河流中，河道和漫滩是合并在一起而且没有明显界限的。在这种情况下，就很难严格定义洪水的发生。因为这个原因，水文学家就将洪水定义为"河流流量的显著增加"。

这两种定义之间的区别是体现在河岸上的，而河岸是人类建造用来保护漫滩的，而人们又把漫滩称作"农田地"。因为课题主要与漫滩淹没有关，所以我们使用前一种定义。

什么是洪水评价？

洪水评价可以分成两个部分：自然科学方面和社会经济方面。第一方面与集水区的自然水文和河道、漫滩的水文特性有关。这些决定了河边地区被淹没的频率、深度以及持续时间。第二个方面指出了在经济方面土地对洪水的敏感度，并与土地利用和洪水的社会认知力有关。

2 洪水的重要性

2.1 洪水——朋友还是敌人

洪水对于人类的影响比其他任何自然灾害都大。在中国，仅 1998 年的洪水就造成 3000 人死亡，影响到数以百万计的人，造成了 200 亿元的损失。在 2000 年，莫桑比克大部分地区被洪水淹没，不计其数的人丧命。因此，通常认为洪水是一种灾害性的自然现象，我们应该努力去控制它，例如修建大坝，是一种对付洪水的一般方法。而为水力发电这种其他目的修建的大坝，防洪被看做第二大作用。

与这种对洪水的消极看法相反，洪水也被看做许多生态系统必备的方面。由于长时间的淹没，像亚马孙、扎伊尔、尼日尔、湄公河、赞比西河许多河流的漫滩形成可以发展其他生产力的湿地生态系统，它反过来又为人类社会所用。洪水提供了自然灌溉，肥沃的漫滩土，形成风景区以及保持生物多样性。

洪水从本质上来说既不好也不坏。如果洪水用来自然灌溉，发展养殖和肥沃土壤，它是好的；如果它淹死人，传播疾病，或者破坏工业、人类财产以及农作物或食物来源，它就是坏的。实际上，由于人类在漫滩居住，强度大、频率低的洪水就是一种大灾害。然而，从生态角度来讲，在沉积物运输和风景开发方面，它又起着很重要的作用。

2.2 洪水过程及其多样性

洪水分析最重要的一步就是了解洪水工程。洪水一般是由于流域上游大量的降水或冰雪融水使得流量超过了河道的容量而造成的。然而，洪水也可能是由当地的降雨或河道的支流引起的。在许多洪水区，洪水过程是顺序进行的。降雨产生最初的地面渗流，饱和后产生地表径流，形成最初的洪水。与此同时，汇入河道的径流也会溢出河道，形成泛滥。如喀麦隆的隆艮洪区。然而，不同的河流洪水形成的原因也是不同的。例如，亚马孙河流域，就是因为地下水位的上涨而形成洪区。洪水的规模一般取决于以下四个因素：

（1）河流的排水面积。
（2）降雨量（或雪融量）。
（3）汇入河流的降水的可能性（与地质和土壤类型有关）。
（4）径流速度（与流域坡度，土地利用以及河道的地貌有关）。

树木和其他浓密的作物的截流作用会有助于降水的下渗，这不仅减少了产流量，还延长了汇流时间。因此，砍伐森林和过度放牧会增加发生洪水的风险。例如，喜马拉雅山区森林的砍伐确实加剧了恒河、孟加拉的布拉马普德拉河下游的洪水的发生。城市化（大面积的屋顶和路面等）和冻结的或牲畜、农业机械造成的坚实的土壤，使得渗透性下降，于是，就有大量降水汇入河道。而在土质渗透性良好的流域降水形成洪水的机会就会大大减小。

在小流域地区，流量会随着降水的发生而迅速变化，因此每年都可能发生多次洪水。而在较大的流域系统，河水流量对降水的反应比较迟缓，因此，很多河流就会有每年一次或两次明显的洪水期。这在世界上那些有明显雨季气候的地区尤其突出，如印度的雨季。该项研究是关于那些洪水为每年一次或两次规律性活动的流域系统的，因为在这些流域中，影响洪水生成的生态及社会因素发生了变化。

任何河流的洪水规模每年都会发生变化，但越大的洪水发生的概率越小。水力学专家已经对洪水进行频率分析，绘制了洪水频率曲线，以确定洪水规模与发生频率的关系。河流上的建筑物，或防洪设置都是按照一定的标准建造的，因此，它们只能承受一定规模的洪水，如百年一遇标准（洪水一旦发生，频率为 100 年一次）。大坝倾向于采用更严格的标准，如可能最大洪水。但是，大坝并不能用来储存特大洪水期间所有的洪水，还需设立泄洪道来排泄多余的洪水以消除对大坝的危害。

3 洪水和生态系统——自然洪水变化的规律对生态系统完整的重要性

河流生态系统包含所有的与河流有关的环境的组成部分（有生物的和非生物的），也包括人类，河道中的水生植物。河流生态系统从源头绵延到大海，包括河岸流域。在河道、洪泛区、湿地、河口和一些河岸地区的地下水主要是依靠淡水的输入。对于能量、事物和物种的流动，河流是一个重要的廊道，而且对地形生态多样性的控制和维持是一个主要的因素。高流速对河道和洪泛区的形态起主要控制作用。它影响着泥沙传输和侵蚀过程。

自然河流的生态系统正适应着自然水文机制依靠洪水来进行交换的这些系统的组成部分，这些组成部分不仅有水，而且有能量营养物质、泥沙和生物有机体。在洪泛区不同的地点，水深、水流形成以及洪水的频率和重现期。在空间和时间上的变化，对多种多样的栖息地和生态多样化起重要的作用，而所有的这些都是由洪水所控制。在世界上，正是由洪水在大多数肥沃的、多种多样的生态系统中创造了河流和洪泛区。

沿海湿地主要依靠淡水输入和河流中所携带的泥沙和营养物质而存在。沿海湿地的生态系统和环境的多样性，是由咸水和淡水的动态边界的波动所引起的。咸水有可能入侵到上游相当长的一段距离，但是边界形成随着水流机制和地形的变化而变化。这些形式不仅影响着植物，而且还影响着动物的行为，例如海洋物种在肥沃的湿地中的分布程度。

4 洪水和自然资源

4.1 洪水和自然资源

自然资源是人类所利用的生态系统的组成部分，例如鱼类、植被、树木和水果。洪泛区的自然资源被定义为两种相关的特征。

特征一：洪水忍耐力——发展成为能够容忍关于洪水环境压力的机制，例如种子和秧苗的生长干扰和饱和土壤中的缺氧问题。

特征二：洪水依靠性——采用这种形式，促使连续的生态系统依靠稳定的季节性洪水。

很明显，在所有的洪泛区生态系统中，在这两种特征之中有一连续体，洪水越有规律，这些系统的洪水依赖性就越强，并且生活在洪泛区的有机体已经使生计策略发展成为依靠有规律的洪水。这种系统能容忍小概率、大洪水事件，而且确实是依靠它们来创造新的生态环境，或是在生长过程中提供特别的生长阶段，例如秧苗移植。另外，很多洪泛区的生态系统能够忍受在两次洪水之间的长时间的干旱。

生活在这个相对稳定的洪泛区的有机体已经采用这种土地利用形式来充分利用这些自然资源生态系统。这种连接返回到社会发展的最初阶段，以集中有机体开始，并且使大多数洪泛区农业得到发展。如那些农业是依靠尼罗河、底格里斯河、幼发拉底河，巴基斯坦的印度河和印度的恒河和中国的长江和黄河。

洪泛区的生态系统以大量不同的方法来支持人类生活区。从机会收获到积极管理，正是大范围的活动描述了洪泛区自然资源的利用。

4.2 退水区域——生产力的关键

漫滩生产力的关键在于退水区域。该区域被洪水覆盖，然后在洪水退回到主航道后便显露出来。退水区域支持多种依靠小地形的生境，并且该区域是一个能够被预计在大多数年里被洪水淹没的地区，这将少于以主要陆地形成洪水事件为特点的地貌漫滩。该地区的生产力是洪水原动力的一个功能。当洪水消退时，高肥料进入季节性沉积物和植物根系区域的保证性可用水，带来了高的生物生产力。

由于地形和水文的较小差别，漫滩以多种自然生境为特点，从地表水系统到草地和森林，支持一系列相似的自然资源使用系统，其中很多分别被认为是"较小"。漫滩自然资源系统的价值是农业、林业、渔业、牧业、狩猎和娱乐这些较小用途的价值的结合，这些形成了漫滩种群生计可持续性的基础。

4.2.1 漫滩渔业

漫滩渔业经常是多产的。因为主要的渠道为鱼类提供了流动性和生境迁移，迁移路线、产卵漫滩和孵卵生境。

对季节性洪水的依靠程度在变化。许多鱼种（常常是经济上重要的）在种群循环和自然洪水事件上表现出牢固关系，这种依靠性可以把摄食和繁殖生境连在一起。为了孵卵，在伊利诺伊河的鲑鲈依靠每年春汛到达被淹没的陆生植被。在尼日尔河的内部三角洲鱼类迁移由洪水量引起并受洪水量限制，而在修建大坝以前，离尼罗河三角洲的现已废止的沿海渔业对洪水运载的营养物质的依靠性很强。

然而，不只是固有的生态生产力重要。渔业的净生产力是初级生产力的一种功能。在洪水消退期间，池塘和 U 字形区域被留在漫滩中，而主要的渠道下降到淡季水平。在干旱季节，鱼类数量在小的地表水域聚集并且很容易捕捉。

4.2.2 河边森林及森林产物

对于所有的天然漫滩体系，河边森林既依赖洪水又要对洪水有防范作用。因为半干旱地区的漫滩森林与洪水出现频率的最小值和地表水的存在时间有关，以及对于形成新生林的位置这类没有规律的事件，防洪看起来仍然是一个决定森林分布的主要因素。在长时间受水淹没的地区，大部分森林物种不会繁盛起来。

森林和林地经常表现出明显的地带性特征，取决于两个方面的影响：一是受洪水出现范围及出现周期影响的局部地形；二是反映受水域影响的植被状况的微小改变。这就导致了物种的产生丰富了森林、林地生态系统。它反过来供养了一大批"次级产物"，与前述产物一起对漫滩群落形成一个很高的累积价值。

在森林生长受水对河岸的适应性限制的干旱和半干旱地区，这些森林的价值显得尤为重要。它是木材和其他产品的唯一来源。但是，即使在森林生长与漫滩无关的湿润地区，漫滩中存在的某些特殊物种同样可以作为本地群落的独特资源。

比较西部亚马孙河的研究，漫滩和旱地森林的利用表明优先受到保护的是漫滩森林。成熟的漫滩森林为当地提供建筑材料和食物，次级漫滩森林对药用植物很重要，湿地森林出产有商业价值的物种。

狩猎不仅仅局限于河漫森林，从生存和收入来源来讲，它对当地群落有极大利益，对许多群落它还有文化和社会方面的重要性。实际上，即使在世界上许多发达国家，娱乐和狩猎这些附属产物逐渐被认为是漫滩最有价值的收益，同时也是湿地保持和重建的最有力证据。

4.2.3 漫滩养牧体系

同其他漫滩群落，草场和它们所依赖的灌溉系统一样变化多端。在无法为森林生长提供足够水量的漫滩或水量太多的地区就会形成草场。

湿地漫滩和干地漫滩是不同的。在干地漫滩存在一个从沼泽草地到腹地干旱草地的连续体，在适当湿润条件下存在森林。在腹地顶级植被是森林的湿地漫滩，草地作为森林的幸存者存在过于湿润或频繁淹没的地区。

为了维持庞大的野生动植物数量，对于不同的气候区已经制订出相应的漫滩放牧体系，范围从温带牧场到干旱的热带漫滩、三角洲。在这些分类中，不同的草地形态存在不同的养牧资源。漫滩放牧体系同自然界野生动植物体系都有季节性迁移，雨季在草地放牧，而旱季则在漫滩地上放牧，这种综合的体制可以使更多的草地用来放牧，从而比只有草地或者漫滩地放牧可以容纳的家畜数目大得多。

除此之外，这种季节性的家畜迁移对于漫滩地或者草地来说在减少过度放牧和生态退化的风险方面也有很多益处。漫滩地对于半干旱和干旱地区的另一个重要影响就是滩地树林或者是岸边树林有大量的营养丰富的植被用来放牧。

人们为了适应季节性迁移而过着游牧生活。他们利用周围环境制定策略，其中包含了灵活性，家畜结构的管理，以及在一些特殊情况下对于环境的管理。在干旱与半干旱地区，牛基本上只吃草本植物，而绵羊、山羊、骆驼则可以吃一些树叶和灌木植物的叶子。这种资源的利用与获取权在干旱季节是被个人或小的集体给包租下来的，而在湿润的季节则一般是公用的。在更湿润的地区，尽管还是有季节性的家畜迁移，但是混合体制的生活（游牧和定居）一般替代了游牧生活。

4.2.4 洪水退后的农业

在湿润地区，漫滩地的农业一般是稻米的种植。亚洲的一些国家，例如孟加拉国、越南、缅甸在半自然状态下也种植深水"浮动稻米"。然而，在特别湿润的地区，人们用水控制系统来控制水的深度，其主要作物是"沼泽稻米"。除了稻米，潮湿的洪水区的主要耕种作物是基于消退洪水的，他们在大堤或者是远离漫滩地或者是在筏子上耕种（亚马孙河流域）。

在干旱地区，特别是在受降雨限制的干旱作物耕种地区，洪水消退后的耕种已经发展成为一种自然的灌溉系统。虽然管理方式变化了，但仍没有超出利用堤岸和明沟进行季节性引水灌溉的方式。

从管理的更高一个层次来说，退洪后的耕种可以看做是洪水管理的一部分。在孟加拉国，农民们基于稻米和多种作物的耕种发展了一套农业系统。这种系统是与季节性洪水"相协调"的。人们在企图控制自然洪水流态来防止其对耕地和城市的破坏失败后，人们正在研究一种新的方法。这种方法是建立在"洪水试验"基础上的，它承认洪水的存在，并用一种复杂的方法计算防止洪水淹没的代价来判断是否需要进行防护，而不是每样事物都进行防护。

4.2.5 生物多样性以及旅游业

除了当地群众直接应用自然资源以外，通过保护以及旅游业的间接利用也是很多洪水依赖系统的重要应用。被保护地区，例如自然公园、自然保护区等用来保护世界上的濒危物种，包括喀麦隆隆艮漫滩地的瓦萨自然公园、博茨瓦纳的奥卡万戈三角洲、巴基斯坦的印度河三角洲红木森林。在这些地区，旅游业可以带来丰厚的收入。然而，落后的农业与野生动植物的保护是相冲突的，通过一些计划例如赞比西河流域的营火计划，当地的群众会从生态以及物种的保护中受益。

山东省污水处理现状概述

赵亭月
山东省水利科学研究院，济南

1 山东省的水污染概况

山东是一个人口大省，"九五"期间在省委、省政府的领导下，经过山东省环境保护工作者的共同努力，较好地完成了山东省环境保护"九五"计划确定的主要任务。山东省环境污染和生态破坏加剧的趋势得到初步控制，部分流域和城市环境质量有所改善，"一控双达标"的任务基本完成，积极推进污染防治和综合整治系统工程，加大环境监督与管理的力度，强化污染物排放总量控制措施。环境目标责任制和城市环境综合整治得到深化，环境污染和生态破坏加剧的趋势总体上得到遏制，重点流域和区域环境质量有所改善。

（1）主要污染物排放总量控制计划全面完成。"九五"期间，在山东省国内生产总值年均增长11%的情况下，全面完成了国家和省政府下达的主要污染物排放总量控制计划。

（2）工业污染源基本实现达标排放。各级政府下达的7795个工业污染源限期治理项目中，有6625个基本实现了主要污染物达标排放，1168个依法停产或关闭。结合经济结构调整，山东省共关停取缔污染严重的"土小"企业4666家，关停万吨左右草浆生产线39条。建设项目坚持先评价后建设，有污染的项目执行"三同时"和"以新带老"规定，基本做到了增产不增污。

（3）流域污染综合治理取得较大进展。组织实施了淮河、海河流域水污染防治规划和小清河、南四湖、东平湖流域污染综合治理总体规划。完成了国家淮河、海河流域水污染防治规划确定的主要污染物排放总量控制任务，初步实现了跨省界断面水质规划目标。

（4）环境法制建设取得了重大进展。为了改善山东省的水环境质量，在"十五"期间，山东省水环境污染防治的主要目标是：工业结构性污染问题得到较好解决，城市生活污水集中处理率达到50%以上，山东省水污染恶化的趋势得到有效控制。

2003年，地表水功能区达标率为44.8%，比上年增加了10%；省控3个湖泊、5座水库氮磷污染程度有所减轻；山东省省控河流污染程度有所减轻。地表水中主要污染物为有机耗氧物质、氨氮和石油类，重金属及其有毒类污染物全部在控制范围内。

2 污水处理的水平及基本措施

2.1 污水处理的水平

山东省的环境保护重点项目截止到 2003 年底,《山东省环境保护"十五"计划》确定的 227 个重点工程项目,已完成的占 42.7%,在建的占 49.3%,未开工的占 7.9%;纳入国家《淮河流域水污染防治"十五"计划》的 225 个项目,已完成的占 25.3%,在建的占 26.7%,未开工的占 48.0%;纳入国家《海河流域水污染防治"十五"计划》的 62 个项目,已完成的占 38.7%,在建的占 32.3%,未开工的占 29.0%;纳入国家《渤海碧海行动计划》的 151 个项目,已完成的占 54.0%,在建的占 37.0%,未开工的占 9.0%;《南水北调东线工程山东段水污染防治总体规划》确定的 217 个项目,已完成的占 29.5%,在建的占 27.2%,未开工的占 43.3%。

流域污染综合治理,山东省政府批准了《南水北调东线工程山东段水污染防治总体规划》《山东省辖淮河流域水污染防治"十五"实施计划》和《山东省辖海河流域水污染防治"十五"实施计划》,省政府召开了山东省重点流域水污染防治工作会议,提出建立严格规范的排污总量控制体系。省政府办公厅印发了《关于进一步加强小清河流域水污染防治工作的意见》,确定到 2012 年实现小清河水体变清目标。

城市污水处理,山东省大、中、小各类城市中,已建成城市污水处理厂 68 个,日污水处理能力 384.8 万 m^3,年污水处理量 7.5 亿 m^3,污水处理厂集中处理率 43.5%,比 2002 年增加 3.2 个百分点。在监测的 42 家城市污水处理厂中,化学需氧量排放达标的 35 家,占监测总数的 83.3%。

2003 年,山东省废水排放总量 24.6 亿 t,比上年增长 6.5%。其中,生活污水排放量 13.0 亿 t,比上年增长 4.7%,占废水排放总量的 52.8%;工业废水排放量 11.6 亿 t,比上年增长 8.7%,占废水排放总量的 47.2%。工业废水排放量中,造纸及纸制品业 3.3 亿 t,与上年基本持平,占工业废水排放量的 30.1%;化学原料及化学制品制造业 1.4 亿 t,比上年增长 15.8%,占工业废水排放量的 12.7%。

废水中主要污染物化学需氧量(COD)排放总量为 82.9 万 t,比上年减少了 3 万 t,下降了 3.5%。其中,生活化学需氧量排放量为 42.3 万 t,比上年下降了 4.4%,占化学需氧量排放总量的 51.0%;工业化学需氧量排放量为 40.6 万 t,比上年下降了 2.5%,占化学需氧量排放总量的 49.0%。工业化学需氧量排放量中,造纸及纸制品业 17.6 万 t,比上年减少了 7 万 t,占工业化学需氧量排放量的 46.8%。由此可以看出,造纸及纸制品业仍是工业化学需氧量排放量的第一大行业。

废水中主要污染物氨氮的排放总量为 7.8 万 t,比上年减少了 0.6 万 t,下降了 7.4%。其中,生活氨氮的排放量为 5.2 万 t,比上年下降了 2.8%,占氨氮排放总量的 68.0%;工业氨氮的排放量为 2.6 万 t,比上年下降了 15.4%,占氨氮排放总量的 32.0%。工业氨氮的排放量中,化学原料及化学制品制造业 0.96 万 t,占工业氨氮排放量的 39.9%。

从以上数据可以看出,虽然污水的排放量较上年有所增加,但污染总量却有很大的下降,证明了山东省的污染总量的排放得到了有效的控制和削减,污水的达标排放得到了很好的控制。

2.2 污水处理的基本措施

污水处理的基本措施有行政措施及技术措施。

2.2.1 行政措施

污水处理的行政措施是，有污染的新建、扩建建设项目，坚持先评价后建设。先评定项目的清洁生产能力，取得排污达标措施。执行"三同时"和"以新带老"规定，在一个区域基本做到增项、增产不增污，增产又减污，大力发展相关高新技术。按照"谁污染谁付费、谁破坏谁治理、谁保护谁受益"的原则，加大企业污染治理资金投入。

对现有工业污染源要实施引导性标准和再提高工程，加大结构性污染治理力度，改善能源结构，推动企业升级换代，努力走出一条科技含量高、经济效益好、资源消耗低、环境污染少、人力资源得到充分发挥的新型工业化路子。加强城市环境基础设施建设，提高城市污水和垃圾集中处理率，实现废物减量化、资源化和无害化。加强农业和农村污染防治。引导农民科学合理施用化肥农药，大力发展生态农业和绿色食品。加快规模化畜禽养殖场污染综合治理，实现养殖废物无害化和资源化。

2.2.2 技术措施

山东省的污水处理技术措施，从方法上基本采用了生物降解法、物理沉淀过滤法、化学氧化-还原法、物理化学联合法等理论。根据不同的污水性质，采用国际先进的组合工艺。

在形式上，对于污染浓度较重的企业实行单独处理达标排放，对于有机废水采用厌氧处理并可回收"沼气"将其资源化。污染浓度小、排放量又较大的企业，对于有机物（COD、BOD）的去除一般采用好氧—活性淤泥法。

城市生活污水一般采用集中建污水处理厂进行集中处理，然后达标排放。对于处理后的微污染水，水中的污染物又以氮、磷类物质为主的污水，在安全的情况下可以进行有指导的农业灌溉，一是利用植物对营养的吸收将水中的污染物吸收，二是为植物提供了营养，变废为宝，完成了污水的最终处理。

湿地系统在山东省将是一个很有开发、利用价值的水处理系统。山东省的自然湖泊较丰富，河口滩涂地较开阔，海岸线的盐碱荒地也较多，这些都是建立湿地系统的有利条件，在海岸线的盐碱荒地建立湿地系统，一是利用自然的荒地系统处理微污染水，二是这些污染水对于海水来讲也是可利用的资源，可以使重盐碱地脱盐形成适合耐盐植物生长的生态系统，优化海岸的荒地环境。

3 污水处理所面临的问题

3.1 自然状况不容乐观

虽然山东省的治污工作取得了很大的进展，但由于我国的国情，一些生产工艺还比较落后及生产成本、治污成本较高，污水的达标排放还不能有效的控制，漏排、私排、偷排，还不能有效、及时、彻底的制止。为此，水污染的问题依然严重。地表水质依然不容乐观。地下水污染由于受土壤的特殊功能的影响，污染效应的滞后性仍然存在。淄河、孝妇河、织女河、小清河等河流沿岸地下水已被严重污染，许多有害成分严重超标。地下水的污染，将在相当长时间内都难以消除。地下水质恶化，给当地居民生活造成极大危害，也使可利用的水资源量减少，加剧了水资源供需矛盾。

黄河干流山东段 3 个断面中，入境断面刘庄水文站水质符合 V 类标准，达不到山东省地表水功能区标准（Ⅲ类）的要求，主要污染物为石油类，其余 2 个断面水质符合Ⅲ类标准，与上年基本持平。

省辖淮河流域水体污染程度较上年有所减轻。优于Ⅳ类水质的断面比率增加 11.1%，劣于 V 类水质的断面比率下降 5.5%。其中，南四湖水系的东渔河、韩庄运河及沂沭河水系的白马河水质有明显的改善。水体中主要污染物仍然为有机耗氧物质、氨氮、挥发酚和石油类。

省辖海、河流域 25 个断面（3 个断面断流），仅有徒骇河申桥断面水质符合Ⅳ类标准、徒骇河毕屯和莘桥断面水质符合 V 类标准，其余断面水质均属劣 V 类，除徒骇河水质有所好转外，其他河流污染仍较严重。水体中主要污染物为有机耗氧物质、氨氮、挥发酚和石油类。与上年相比，水质无明显变化。

小清河干流 7 个断面，除源头睦里庄断面水质符合Ⅱ类标准外，其余 6 个断面水质均低于劣 V 类标准。水体中主要污染物为有机耗氧物质、氨氮和石油类。与上年相比，水质总体上无明显变化。

3 座湖泊中，东平湖、南四湖等 11 个测点水质均低于劣 V 类标准。

5 座大型水库中，峡山水库 3 个测点水质均符合Ⅲ类标准，崂山水库和门楼水库 6 个测点水质低于劣 V 类标准。

山东省湖泊、水库主要超标项目为总氮，根据湖泊水库营养化程度评价结果，崂山水库、门楼水库、峡山水库属中营养，南四湖属轻度富营养。

3.2 任务艰巨

主要污染物排放总量控制计划《山东省环境保护"十五"计划》确定的山东省主要污染物排放总量控制指标，要完成 2005 年总量控制计划任务十分艰巨。在 2003 年排放总量的基础上，化学需氧量排放量需削减 5.51 万 t，削减率 6.6%；SO_2 排放量需削减 31.6 万 t，削减率 17.2%；烟尘排放量需削减 5.4 万 t，削减率 8.7%；工业粉尘排放量需削减 18.5 万 t，削减率 25.0%。这对于山东省的污染治理来说是一个非常严峻的数据。

3.3 污染的转移

随着城市对污染治理法规的完善，执法力度的加大，一些污染较大，产值利润相对又较小的企业在逐渐向小城镇转移。城市的污水通过河流大多流向农村的土地渗入到地下水。由于山东省乃至全国的农村状况，人们的环境意识还较淡薄，利益的驱动，地方主义的概念，使得对于污水的控制与治理更加无序化、分散化。

再就是农村科学技术的落后。目前，由于不合理地使用污水进行灌溉，已经造成部分农田土壤、作物及地下水的严重污染，污水灌溉引发的农田污染事故不断发生，对农村水环境构成威胁。污水灌溉已成为山东省农村水环境恶化的主要成因之一。由于在小城镇污废水总体处理率低，排放水质超标，污灌前普遍缺少必要的污水预处理措施，大部分污废水没有达标或未经处理则直接用于灌溉。由于近几年的淡水短缺，污水灌溉面积盲目发展，监控、管理体系不健全，河道灌溉功能退化。

在随着农村近几年来经济的发展，集约化养殖成了支撑农村经济的主要力量，但养殖场管理不善也对农村水环境造成了严重的威胁。以2000年为例，全国畜禽粪便的年排放量超过27亿t，已超过全国工业废渣和城市生活废弃物排放量之和。由山东省生态现状调查与规模化畜禽养殖业污染调查发现，2000年末山东省畜牧业规模饲养户达26.2万个，生猪、牛、羊存栏分别为2660.3万头、1008.6万头、2784.7万只，家禽存养数5.62亿只。畜禽粪便含有较高的COD、BOD、SS，这些本应是农业较好的有机肥料物质，由于管理不善，随意排放或随降雨径流污染水体。

综上所述，近几年来山东省的污水处理尽管取得了长足的进展，但目前来看，山东省的污水治理任务仍然相当严峻。

4 污水的合理利用

人们使用了水，在造成污染的同时也获得了很大的经济利益，推动了社会经济的发展。污水的治理是利用科学的技术，推动清洁生产，一是把污染降到最小值防治水污染，二是合理地处理污水。为此污水的再利用问题也将是一个科学的、积极的污水处理方向。

事实上，在自然生态中维持生态平衡的诸多因素中，每一个元素（物质），都有它的特殊功能，所不同的只是数量的限制和出现的时间、地点。例如，污水中的元素N、P、K是植物生长的三大主要元素，有些元素还可能是植物生长的调节剂。所以说，每个事物都有它的双重性。再如，化肥一方面是为作物提供了营养物质，但另一个角度它也污染了土壤并间接地污染了地下水。为此，可以说将污水正确分类，因地制宜地对它再利用或者通过利用对其合理分配，使污水资源化，既是一项节水措施，也是一项治污措施，兼有经济效益、社会效益和生态效益。

4.1 污水处理的生态工程

污水生态工程主要是湿地及污水的安全灌溉，是充分利用土地和植被的自然净化生态功能截留、净化、利用、转移污水中的N、P及有机物，加强N、P等物质在陆地、湿地生态系统的循环，从而减少面源污染对水体的污染。

目前，我国污水灌溉农田面积330余万hm^2，占全国总灌溉农田面积的7.3%，主要分布在我国北方水资源严重短缺的海、辽、黄、淮四大流域，约占全国污水灌溉面积的85%，大部分集中分布在相应大中城市的近郊区或工矿区。山东省也是大量利用污水灌溉的省份之一，据测算估计约有30万hm^2。

4.2 分质供水

中水利用是根据用水户对水质的需求，将工业和生活的污废水经过处理，达到某种标准后再利用。

在工业工艺用水中，不同的工艺对水质的要求也相差很大，为此可以使用逆流程的分质供水，即下一个工序流程的废水经处理后用到上一个工序流程。例如，山东省的高唐银河纸业集团，将纸机废水经气浮技术处理后用到制浆工序，使纸机白水的利用率达到 90%以上。

4.3 景观用水

在城市近郊可以利用人工景观水体的自然氧化、水生观赏植物的功能对微污染水质进一步深度处理。

5 结语

人们利用了水，在取得需要、获得利益的同时污染了水，这是必然的，也是人类生存必需的。从自然的角度看也是符合生态规律的。但问题是人们必须要科学地掌握好尺度。利用了环境，就必须回报环境、保护环境。人们必须认真地研究如何合理用水、节约用水，必须要认真地做好污水的处理工作，保证水资源的可持续利用。保护好水环境，保护好生存环境，保护好我们的家园。

山东省农业用水价格改革方案及供水协会建设规划研究

刘建强
山东省水利科学研究院，济南

1 农业用水价格改革方案

水的问题已引起全社会的普遍关注，水利事业正处于一个重要的转型期，当今社会对水的认识达到了很高的程度，中央领导已把水安全放在水、石油、粮食三大安全问题的第一位，国民经济和社会发展"十五"计划纲要把水利摆在基础设施的首位，列交通、能源之前。山东省委、省政府指出水是21世纪关系山东省国民经济发展的重大问题。水利事业正面临着千载难逢的发展机遇，也面临着前所未有的挑战。工程水利向资源水利的转变，从很大程度上说是从性质上的转变，它强调水的综合治理，人和水的和谐相处，以水定发展，以水定国民经济和产业结构的发展。长期以来，水利主要是围绕着农业转，传统水利向现代水利、可持续发展水利转变，主要体现在由农业水利向全社会大水利的转变。但是随着经济的发展，水资源供需矛盾日益尖锐。特别是长期以来由于水价关系未理顺，水资源浪费现象极为严重。由于现行水价标准低，水费计收情况较差，没有把水费纳入市场经济体制下的价格体系管理，缺乏市场调节机能；没有形成对水利工程供水的相应补偿机制，严重制约了水管单位的经营管理能力。因此，规范水价管理，建立合理的水价制度，探讨农业用水价格改革机制，对合理使用水资源有着极其重要的意义。按照市场经济规律，运用经济杠杆，因地制宜地进行水价改革，是促进水资源优化配置，节约用水，保证工程正常运行的根本，也是发展壮大水利基础设施和基础产业所需。因此，采取有效措施，深化水价改革，适时建立合理的水价形成机制，是促进水利事业发展的关键环节。

1.1 农业用水价格改革的基本原则和思路

1.1.1 农业用水价格改革的基本原则

在充分考虑农民承受能力的基础上，通过建立科学合理的农业水价形成机制，提高农民的节水意识，促进农业灌溉方式的转变，达到节约用水的目的。改革农业水价必须与改革农业供水体制和改造农业灌区、改善计量手段等统筹安排、配套实施。

1.1.2 农业水价改革的基本思路

（1）深化农业用水价格管理体制改革。

现行农业用水仍沿用历史形成的价格管理体制，即农业灌区支渠以上的供水工程由水行政主管部门管理，以支渠进水口为定价计量点，水价由省价格主管部门会同省水行政主管部门制定或调整；支渠进水口到支渠出水口供水工程由水行政主管部门管理，但没有出水口的水价标准；支渠出水口到田间地头的供水工程由乡镇或行政村管理，其加价标准由乡镇或农民用水者协会确定。各灌区大都实行综合估算、按亩收费的办法。这种灌区上游水价由政府管理、下游水价由乡镇或农村自行确定的水价管理体制，从客观上难以约束乱加价和搭车收费行为，为此，必须改革和完善现行的农业供水价格管理体制。一是将农业供水各环节水价均纳入政府价格管理范围，实行政府定价和政府指导价两种价格形式。国有水利供水工程（包括引黄灌区、引库灌区、引河、湖灌区、排灌站等）向农业供水价格实行政府定价，大型水利工程由省价格主管部门会同省水行政主管部门制定，中型水利工程由市价格主管部门会同市水行政主管部门制定，定价计量点由支渠进水口改为支渠出水口。已实行产权改革的、民办、民营中小型水利工程供农业用水价格实行政府指导价，由市价格主管部门会同水行政主管部门制定基准价格和浮动幅度，具体水价标准由供水工程经营者与用水户在指导价范围内协商确定。二是加强对乡镇及以下供水环节的成本核定和水价管理，通过水利工程取水的自流灌区，可实行在省核定的支渠出水口供水价格的基础上加末级渠系的损耗和维护费用的定价模式，末级渠系的损耗及维护费用标准由各市价格主管部门会同水行政主管部门制定，末级渠系维护费按照补偿乡镇及以下供水渠系维护管理合理成本的原则从严核定。

（2）充分考虑农民的支付能力，逐步解决农业水价与供水成本倒挂问题。

在清理整顿水价秩序，取消农业供水中间环节乱加价和搭车收费的基础上，充分考虑农民的承受能力，分步将农业水价提高到供水成本的水平。对大型高扬程提水、机电井灌区及其他成本高的水利工程（如跨流域调水工程），要采取提价和扶持政策相结合的办法，适当解决供水成本与价格倒挂问题，减轻水价上涨压力。

（3）健全水费计收和管理制度，降低农业供水成本，减轻农民水费负担。

一是认真贯彻《水利产业政策》，将水利工程水费严格纳入商品价格管理并加强监督检查，由水管单位作为经营收入直接收取，防止水费收取中的加价、截留、挪用等行为。二是健全水费收取制度，增加透明度。在大型灌区要推行"统一票据、明码标价、开票到户"的办法，实行账务公开，在村组将放水时间、水量、水价、水费予以公布，接受群众监督。三是约束供水成本，完善水费使用办法。水管单位要加强内部管理，杜绝不合理的开支；要按财务规定提取折旧和修理费，专项用于水利工程设施的维修、更新和改造，不得用于发放奖金等非生产性开支。

1.2 农业水价改革的具体实施办法

1.2.1 结合农村治乱减负工作，整顿水价秩序

加强农业水价管理，彻底取消不合法的加价和收费，特别是不得趁改革水价之机，巧立名目加价收费。供水单位与农民的结算价格由规定的水利工程供农业水价加末级渠系的损耗、维护费用组成，除此之外，任何单位不得另行加价、收费；同时，各级物价、水利

部门要开展农业供水成本测算和审核工作，规范正常的供水成本费用，提高水价构成的透明度。各地要加强对整顿和规范农业水价工作的领导，各市物价部门要会同水利部门，集中组织力量对本辖区内农业用水中的乱加价、乱收费进行调查清理，并做好农业供水成本核算工作，为调整水价打好基础。

1.2.2 在清理整顿的基础上，逐步调整农业水价

在清理整顿和严格审核农业供水成本的基础上，充分考虑农民的实际承受能力，制定农业水价调整方案。按照国家计委农业水价原则上三年内逐步调整到位和山东省政府鲁政发[2001]50号文件确定的"2002年前，农业供水中的粮食作物用水价格要达到保本水平，经济作物用水价格要达到保本微利"原则，进行农业用水价格的调整。

1.2.3 农业用水价格水平总体安排

根据以上原则，考虑农业供水成本和农民承受能力等因素，山东省各类水利工程供农业用水价格水平安排意见是：大中型水库灌区 $0.12\sim0.15$ 元/m^3，提高 $5\sim8$ 分；引黄灌区 $0.11\sim0.14$ 元/m^3，提高 $5.4\sim8.4$ 分；引河、湖灌区 $0.11\sim0.13$ 元/m^3，提高 $5\sim7$ 分。民办、民营中小型水利工程向农业供水价格中准价格及浮动幅度由各市物价部门会同水利部门根据工程投资等情况，参照上述水平确定。

1.2.4 加大对农业水价改革的宣传力度

农业水价改革的根本目的是促进农业节水。各地要加强宣传，提高全社会特别是农民对改革农业水价必要性和紧迫性的认识，保证改革顺利进行。

1.2.5 加强农业水价改革的研究和探索

山东省在农用供水体制和价格管理体制方面还存在不少问题，农业节水的潜力很大，结合农业供水体制改革，必须进一步研究利用价格杠杆促进农业节水问题。一是适当引入农业供水地区差价和季节差价。可根据水资源条件和供水工程情况实行分区域或分灌区定价。利用浮动价格机制，加大水价的激励和约束作用，缓解水资源紧缺的矛盾。二是创造条件逐步实行超定额用水累进加价。在目前农业用水计量设施较为落后的条件下，可按照水的自然流程和灌溉区域，以打破行政区划的自然村、组或农民用水合作组织为水量计量和水费计收对象，分别制定基本用水定额，定额内用水价格不提高或小幅提高，超定额用水可较大幅度提价并实行累进加价，促使购水单位采取措施合理配水、控制损耗，促进农户节约用水。三是为促进水资源的合理分配和水利工程的稳定运行，可在农业水价中实行计量水价和容量水价相结合的两部制水价。此种计量方式已在山东省的部分地区应用，效果良好。

1.3 农业水价改革的配套措施

1.3.1 推行农业供水产权制度改革等体制创新，降低农业供水的中间交易成本

从长远来看，农业节水的根本出路在于农村经济实现规模经营，水管部门直接供水管理到农户，实行按水量计价。近期内要重点理顺乡镇以下供水环节的供水体制，可考虑推行以下几类体制替代基层行政组织管水。一是国有水管单位管理范围延伸，按水的流程直接管理、收费到自然村、组或农户。二是逐步建立用水者协会等形式的农民用水合作组织，让农民自己参与管水，负责末级渠系的水量分配、水费收取和渠道维修等工作，实行民主管理。三是实行小型水利产权制度改革，对乡村输水渠道进行公开租赁或承包，将农业灌

溉管理经营权移交给农民,明确承包者对农户的最终水费标准,通过管理权限、职责和利益的挂钩,调动农民维修渠道、节约用水的积极性,理顺管理体制。农业供水中间环节和费用将大大减少,可有效降低农户实际水费。

1.3.2 总结、示范和推广有效的农业节水技术

认真总结现有的节水技术和经验,加强节水技术交流,结合各地的自然和经济条件,大力推广实用、先进的节水技术和灌溉方式。一是推广渠道防渗和管道输水技术,减少输水过程中的损失。二是休整输水渠道,平整耕地,改进田间节水配套措施。三是推行科学适时、适量灌溉,根据农作物生长周期和品种特性安排灌溉,减少灌溉的盲目性。四是有条件的地区可因地制宜推广喷灌、滴灌等田间节水措施。

1.3.3 加大渠系改造和节水灌溉投入,提供节约用水的设施和技术保障

采取"提投并举"的方针,在提高水价的同时加大对农田水利和节水灌溉工程的资金投入,并适当引入多元化投资机制,以加大农业灌溉和计量设施改造,逐步实现渠道衬砌和计量收费。水管单位也要将增加的水费收入主要用于供水设施的维修和养护。

1.3.4 调整农业水旱结构,提高农业对水价的承载能力

实行水旱互补的方针,重视发展旱作农业,加强旱作农业示范基地建设,大力推广旱作农业技术,因地制宜调整水旱种植面积并改变灌溉方式。贫水地区应少种高耗水作物,多种节水作物。通过种植结构的优化调整,实现农业区域分工互补,缓解水资源时空分布不均的矛盾。

1.4 结语

上文根据农村经济发展状况和水利产业政策的要求,在充分考虑农民承受能力的前提下,提出了适度提高农业用水价格,促进农村节约用水,合理增加供水单位收入,增强水管单位对工程的运行和维护能力,加强测水量水设施建设,提高供水质量,促进水资源合理利用的水价管理模式。特别是对农业水价改革提出了累进制水价加价制度及两部制水价制度,经在山东省部分水利工程中应用,效果良好,具有重要的推广应用价值。

2 供水协会建设规划方案研究

2.1 农村水利协会的主要内容

2.1.1 农村水利协会的内涵

农村水利协会,是由灌溉供水区、饮水供水区的受益农民自愿参加组成的群众性用水管理组织。经当地民政部门登记注册后,具有独立法人资格,实行独立核算,自负盈亏,实现经济自立。它与其上一级供水组织(供水公司)共同建立了符合市场经济机制的供水和用水的供求关系,实现商品化供水和用水。农民用水者协会负责所辖灌溉系统及供水系统的管理和运行,保证灌溉工程、供水工程资产的保值和增值,同时向供水公司交纳灌溉水费。农民用水者协会由用水小组和用水者协会会员组成。

(1)用水小组:是将供水受益区内的用水户按地形条件,水文边界,街道分布情况,同时兼顾村、村民小组边界划分成的若干个用水单位。每个用水小组设立用水者代表,即

是该组的管水员,负责执行用水者代表大会作出的决定,管理本组的灌溉设施、供水设施和用水工作,向本组用水户收取灌溉水费和饮用水费,并负责本组的其他日常工作。

(2) 用水者协会会员:是指在协会范围内的受益农户户主,由用水户自愿申请加入,并经用水者协会批准。其权利和义务主要是:参加本组大会,有灌溉用水、人畜用水的权利和按时交纳水费的义务,并根据要求参加保护、维修协会管理范围内的灌排设施和供水工程设施。

2.1.2 农村水利协会的特点

(1) 农村水利协会是具有独立法人地位的社会团体组织,在民政部门注册登记,取得合法地位,依照国家的法律、法规及协会章程运作,并对管理设施、财产和灌溉服务、人畜吃水供水服务负法律责任。

(2) 农村水利协会是不以营利为目的的服务性经济实体。协会向用水户提供灌溉供水和灌排技术服务以及人畜吃水工程的技术服务,维护农民用水户的相关利益,不以营利为目的。同时,它又是一个经济实体,它与骨干工程的灌区(或供水厂)专管机构(供水公司)是特殊商品——灌溉用水或人畜用水的买卖关系而不是上下级关系。其水费收入除支付买水的水源费以外,将全部用于协会管辖范围内的工程维护、运行、管理的人员开支等,任何其他组织或个人不得挤占挪用。

(3) 对于灌溉工程,农村水利协会通常按渠系流域组建;对于饮水工程,用水者协会通常按行政村或村民小组组建。它们均不受当地行政机构的干预。

(4) 农村水利协会是农村从事灌溉服务或人畜用水服务的民主组织,用水户按自己的意愿民主选举协会的组织机构,内部事物通过会员代表大会以民主协商的方式确定,活动受到用水户的广泛监督。

(5) 农村水利协会接受政府业务主管部门(水利管理机构)或灌区专管机构的服务监督和指导。

2.1.3 农村水利协会的组建程序

农村水利协会是由供水受益区农民自愿组成,其组建程序如下:①成立领导小组;②宣传发动;③划分协会和成立协会筹备组;④安排协会办公场所和设施;⑤组织用水户入会,摸清协会基本情况;⑥划分用水组;⑦用水者代表大会召开前的准备工作;⑧召开用水者代表大会,成立用水者协会;⑨资产移交,产权明晰;⑩注册登记;⑪制定协会规章制度;⑫组织验收。

2.2 农村水利协会规划的指导思想、总体思路与原则

2.2.1 规划方案指导思想

以邓小平理论和"三个代表"重要思想为指导,全面贯彻党的十六大和十七大会议精神,坚持以人为本,树立全面、协调、可持续的发展观,适应全面建设小康社会的总体要求,以提高农村水利工程供水安全为目标,结合水利部出台的《小型农村水利工程管理体制改革实施意见》,统筹规划,突出重点,加快山东省小型农村水利工程管理体制改革进入法制化、规范化和科学化的轨道。力争在较短的时间内,以明晰工程所有权为核心,全面完成现有小型农村水利工程管理体制改革,逐步建立适应社会主义市场经济体制和农村经济发展要求的工程管理体制和良性运行机制。改善农民的生存环境、生活和生产条件,

促进农村脱贫致富，尽快达到全面建设小康社会的总体要求。

2.2.2 规划方案总体思路

规划农村水利协会发展的总体思路是：按适应全面建设小康社会的总体要求和《小型农村水利工程管理体制改革实施意见》，以提高农村灌溉用水方便条件、实现饮水供水安全及时为目标，统筹规划，分步实施，力争在3~5年内，以明晰工程所有权为核心，全面完成现有小型农村水利工程管理体制改革，逐步建立适应社会主义市场经济体制和农村经济发展要求的工程管理体制和良性运行机制。按照"先急后缓、先重后轻、分阶段、分步骤实施"的原则，优先解决对农民生活和身心健康影响较大的供水工程管理改革问题。在确保工程安全的前提下，充分引入市场竞争机制；坚持责、权、利相统一，实行"谁投资、谁所有，谁受益、谁负担"；坚持政府扶持与农民自主兴办相结合，鼓励社会各界参与工程建设和管理；坚持因地制宜、民主决策，积极稳妥地推进改革。要明确不同类型小型农村水利工程的管理体制和产权归属。农户自用为主的小微型工程，产权归个人所有，由县级水行政主管部门统一监制，乡镇人民政府核发产权证；对受益户较多的工程，组建用水合作组织管理，国家补助形成的资产划归用水合作组织；对村镇集中供水工程，视工程规模组建工程管理委员会或用水合作组织管理；对经营性的工程，组建法人实体，实行企业化运作，国家补助形成的资产可由乡镇委托水管站等组织持股经营，也可卖给个人经营。根据市场经济的特点，引入了科学的运行机制，以合同为纽带，通过用水合作组织管理、承包、租赁、股份合作、拍卖等形式，把小型农村水利工程管理体制纳入法制化轨道。严格规定涉及地方人民群众生命财产安全的工程的所有权不能拍卖。同时，要按照市场经济的要求，切实加强组织领导，规范改革运作程序。各级政府要继续加强引导、扶持、服务和监督职能，研究制定有关的政策和法规。各级水行政主管部门要在政府统一领导下，密切配合计划、财政、物价、国土、农业等部门严格执行国家水法规和有关政策，在对工程现状进行调查摸底和搞好宣传发动的基础上，按照划分范围、界定所有权、资产评估、确定方案、公开招标、签订合同、报批备案、建档立卡等步骤组织实施，规范化操作。

2.2.3 规划原则

要建成一个自主管理、衷心为用户服务、人员配备合适、规章制度健全、具备初步工程设施条件、能够受到广大用水户支持的农村水利协会，需要坚持的原则是：

（1）组织建设与思想建设并举原则：用水者协会的组建不仅是一个组织机构上的变化，也是一个思想观念转变的过程。要将过去主要依赖政府改变为完全由用水户及他们自己的代表——用水者协会自主管理供水，以及维护工程，从而增强农民用水户的主人翁意识，积极参与灌溉管理，改善供水条件。因此，在整个组建工作中必须反复向协会人员、用水者代表和广大用水户宣传这一基本观念，使之家喻户晓、深入人心，并坚持贯彻到组建过程中去。

（2）自愿、公开、民主的原则：农民用水者协会的组建必须遵循自愿组织的原则，政府不能进行行政干预，应做好用水者协会的宣传、发动、培训和扶持工作，使农民自愿加入农民用水者协会。

（3）要有相对独立的水源，水资源统一管理的原则：在农民用水者协会的运行管理中，供水公司或供水单位与农民用水者协会每年都要签订供水合同，供水公司或供水单位应按供水合同的要求向农民用水者协会供水，因此，供水公司或供水单位必须有相对独立的、有使用权的水源。另外，水资源的使用应遵守统一使用与管理的原则，任何用水户、用水小组的用水必须通过用水者协会向供水公司或供水单位申请，并经供水公司同意后，才能实施。

（4）因地制宜的原则：农民用水者协会的管理模式一般有供水公司—农民用水者协会、供水单位—农民用水者协会两种形式。在农民用水者协会的建立中，要因地制宜，不强求统一，不搞"一刀切"。有条件的供水单位可经过改制组建供水公司，没有条件或条件不成熟的仍按供水单位管理。在灌溉工程方面，对灌区各支渠、斗渠，有条件的可成立农民用水者协会，暂时不具备条件的可采用其他方式或仍按原方式管理。在饮水工程方面，对于联村供水工程或乡镇集中供水工程，有条件的要积极推广应用乡镇供水模式或县级供水模式；对于单村饮水工程，可以进行先试点，后推广。

（5）依法建立的原则：供水公司和农民用水者协会都具有法律地位，因此，其建立必须遵守国家现行的法律法规。供水公司要到当地工商行政管理部门注册登记，用水者协会要到当地民政部门登记并经过批准。

（6）经济自立的原则：经济自立是农村水利协会的核心与目的，因此，从供水公司和农民用水者协会建立开始就要把握这一原则。运行上自主经营，财务上实行独立核算，有专（兼）职财务人员，厉行节约。

（7）水文边界与行政区划相结合的原则：农民用水者协会中用水小组的划分，在考虑地形地貌、水文边界及街道分布的前提下，尽量与行政村、自然村、生产组相结合，以方便管理。

2.3 规划方案

2.3.1 在灌溉供水工程方面

目前山东省灌区工程基本分为四大类，一是利用地表水灌溉的渠灌区。二是利用地下水灌溉的井灌区。三是利用地表水和地下水联合灌溉的井渠结合灌区（以下简称井渠灌区）。四是以水库为水源的自流灌区。综合各地已建设的农村水利协会的做法，本次规划不同类型的灌区，采用的管理模式如下：

（1）渠灌区用水者协会管理模式：规划在渠灌区采用供水单位加用水者协会的形式，即"供水单位+用水者协会+用水组+用水户"的管理模式。有的试点区采取供水公司加用水者协会形式，即"供水公司+用水者协会+用水组+用水户"的管理模式。供水单位按照"事业性质，企业化管理"的方式，开展各项管理业务。在财务上单独建账，实行独立核算，自负盈亏，自我维持，建立良性运行机制。供水单位不同于按《公司法》组建并到工商行政管理机关进行企业法人登记的供水公司，它只需政府主管部门批准，即成为具有独立法人资格的事业法人。

在渠灌区，规划按独立水文边界组建用水者协会，以免与周边地区发生用水或排水矛盾。在用水者协会试点区内，用水者协会基本上要以一条斗渠或支渠为控制范围，灌溉面积大约在 10 000 亩，以便取得试点经验后加以推广。

（2）井灌区用水者协会管理模式：山东省是机井建设大省，井灌区面积较大。规划井灌区一般以一眼机井为基本单元组建用水小组，由农民推选用水者代表；以一个电力变压器控制面积为单元，成立用水者协会。也可以自然村或行政村成立用水者协会，通过民主选举推荐用水者协会主席和执委会成员。在井灌区一般不成立供水公司，而是采用管水、用水一体化的农民用水者协会，即用水者协会的组织管理模式为"用水者协会+用水小组+用水户"。井灌区用水者协会，主要是对计划用水进行督促，协助政府加强水资源管理，指导用水小组开展节水灌溉和技术服务。各用水组为独立核算的经济组织。

规划在井灌区成立的用水者协会，大田灌溉面积一般为2 000~5 000亩，不宜过大，以便管理。

（3）井渠灌区用水者协会管理模式：在山东省的引黄区域内，为了提高灌溉用水保证程度而采取了"双保险"措施。骨干渠系作为灌区主要的灌溉工程，负责全灌区的输水和配水，实行统一规划，统一设计。机井作为灌区的辅助灌溉设施，每眼井只控制灌区的部分灌溉面积。以渠灌为主、井灌为辅的"井渠灌区"，供水公司和用水者协会的组建及管理模式与渠灌区基本相同。在该类灌区，规划用水者协会一般是以斗渠或支渠为控制范围，按独立的水文边界组建。用水组一般将单井控制范围内的用水户划分在同一个用水组内。灌溉用水的原则是"先地表水，后地下水"，即在实施灌溉中，先用供水单位（公司）提供的地表水，当供水单位（公司）的水源无保证时，则使用井水灌溉。水费核算分两部分，即以供水单位（公司）为水源灌溉的水价和以井水单独灌溉的水价。用水者协会使用供水单位（公司）提供的灌溉水时，用水户水费按用水者协会核算的水价交费，用水者协会向供水单位（公司）上交水费。用水户使用机井灌溉时，按井灌核算的水价向用水者协会交纳水费。

（4）自流灌区用水者协会管理模式：根据水库自流灌区的特点，规划该种类型的灌区用水者协会可按以下模式组建。农民用水者协会一般应按支渠为单位组建，而不考虑行政区划，即每个支渠建立一个农民用水者协会。当支渠控制的灌溉面积较大时（如万亩以上），可将同一支渠划分为2个或2个以上农民用水者协会。本着管理方便的原则，农民用水者协会可打破村的界限，也可保持村的界限，用水小组一般按斗渠为单位划分。当斗渠控制的灌溉面积较大时，也可将一条斗渠划分为2个或2个以上用水小组。

2.3.2 在农村饮水工程方面

根据山东省的地形情况大体分为三个区域进行农村水利协会规划。①胶东半岛低山丘陵区：包括济南市、淄博市、潍坊市、烟台市、威海市和日照市六个市；②鲁西北、鲁西南黄泛平原区：包括东营市、滨州市、德州市、聊城市、济宁市和菏泽市六个市；③鲁中南山丘区：包括泰安市、莱芜市、临沂市和枣庄市四个市。三个分区应因地制宜考虑农村水利协会的规划方案，既要保证饮水工程的长期发挥效益，又要考虑管理方便，符合当地的实际情况。

1）胶东半岛低山丘陵区

胶东半岛低山丘陵区，区内经济状况相对较好，个体、集体经济发展迅速，经济实力相对较强。本着集中连片、突出重点、可持续发展的原则，在实施农村水利协会建设规划时，应采取如下几种模式：

（1）集中供水模式：在地下水水质较好的地区集中开辟供水水源地，可以以乡镇驻地为中心向四周辐射，也可以多个密集村庄统一联网集中成片解决，形成网络化集中供水体系，采取加压泵站直供或利用高位水池自流供水的方式；实行供水公司+乡镇供水协会+用水者协会+用水组+用水户的管理模式。

（2）联村供水模式：在水源水量水质满足要求的条件下，积极发展联村供水模式，以节省工程投资。充分利用丘陵山区地形的特点，修建高位水池来实施自流供水的方式，采用供水自动控制装置，方便管理和运行。实行乡镇供水协会+用水者协会+用水组+用水户的管理模式。

（3）单村供水模式：对于无法实施联村供水且项目村具有一定的经济实力的条件下，可以采取单村供水模式。实行用水者协会+用水组+用水户的管理模式或承包、租赁的管理模式。

2）鲁西北、鲁西南黄泛平原区

鲁西北、鲁西南黄泛平原区主要为水质性缺水地区，一般为浅层淡水—中层咸水—深层淡水，随着地下水位的逐年下降，浅层淡水越来越少，并且越来越咸，深层淡水含氟量高，对人体健康危害较大，不能饮用。在这部分区域，本着以集中联片供水为主，坚持因地制宜、分类指导、可持续发展的原则，在有条件的区域，以黄河水为水源，利用黄河水，兴建地表水拦蓄工程，如平原水库、蓄水坑塘等，借鉴沾化、无棣经验，建立集中供水工程，实行供水公司+乡镇供水协会+用水者协会+用水组+用水户的管理模式。具备联村供水条件的尽量采用联村供水的方式，采用乡镇供水协会+用水者协会+用水组+用水户的管理模式。有向城镇供水的工程要附带解决附近农村的饮水安全问题，向城乡一体化的方向发展，根据工程规模，采取统一集中管理、股份制等管理方式。

3）鲁中南山丘区

鲁中南山丘区为山东省的石灰岩山区和山前冲积平原区，为水源性缺水区域，再加上经济条件差、供水难度大，该区域应坚持以小型供水工程为主、因地制宜和可持续性发展的原则。根据当地的实际条件，采用乡镇集中供水、联村供水、单村供水和水窖供水等方式。

（1）乡镇集中供水模式：在地下水水质丰富的区域打大口井和深井，作为饮水工程的水源，以乡镇驻地为中心向四周村庄辐射，形成乡镇集中供水体系，建多个高位蓄水池，采取自流供水的方式，实行乡镇供水协会+用水者协会+用水组+用水户的管理模式。

（2）联村供水模式：在山丘区，积极推广应用联村供水模式。以中心村庄为主，带动周围的小村庄统一联网集中成片解决饮水问题，水源以浅层地下水或深层地下水为主，建多个高位蓄水池，采取自流供水的方式，实行乡镇供水协会+用水者协会+用水组+用水户的管理模式。

（3）单村供水模式：对于项目村位置偏僻，有一定的地表水源时，可采取单村供水模式，建一个高位蓄水池，采取自流供水的方式，实行用水者协会+用水组+用水户的管理模式或承包、租赁的管理模式。

（4）水窖供水模式：在项目村地表水、地下水缺乏的区域，可以根据雨季地表水丰富的特点，采用水窖供水模式解决饮水安全问题。每年在汛初和汛末水源充沛时灌水两次，即可满足4口之家一年的生活用水。水窖的大小以40~60 m^3 为宜，采取自流或手压泵取水的方式供水，实行用水者协会+用水组+用水户的管理模式。

2.4 规划结果

2.4.1 建设规模

根据上述规划方案，本次规划山东省农村水利协会的发展规模是：

(1) 2006—2010年在山东省范围内,每年发展农村水利协会300个左右,五年内共建设1 500个左右。

(2) 2011—2015年在山东省范围内,每年建设农村水利协会800个左右,五年内共建设4 000个左右(表1)。

表1 不同阶段建设山东省农村水利协会发展数量规划

年度	每年建设农村水利协会数量/个				总计/个	备注
	灌溉协会	供水协会	小型灌溉工程管理联合体与合作组织	小计		
2006—2010年	120	140	40	300	1 500	"十一五"期间
2011—2015年	350	350	100	800	4 000	"十二五"期间

2.4.2 投资估算

根据上述规划方案,结合山东省已建农村水利协会的具体情况和运行管理情况,经调查分析,估算出不同发展阶段发展农村水利协会的投资(表2)。

表2 不同阶段山东省发展农村水利协会投资估算

年度	每年建设农村水利协会投资/万元				总计/万元	备注
	灌溉协会	供水协会	小型灌溉工程管理联合体与合作组织	小计		
2006—2010年	3 000	4 200	800	8 000	40 000	"十一五"期间
2011—2015年	10 500	10 500	2 000	23 000	115 000	"十二五"期间

2.4.3 资金筹措

农村水利协会建设是一项公益性事业,需要政府投资引导和受益群众共同投资进行建设,同时要引导农民投工、投劳。由表2可知,本次规划山东省2006—2010年发展农村水利协会总投资为40 000万元,每年需投资8 000万元;2011—2015年发展农村水利协会总投资为115 000万元,每年需投资23 000万元。在资金筹措方式方面,根据有关要求,按照省级财政补助,地方财政配套和受益群众投资与投劳、投工等办法,进行筹措资金。

3 结语

长期以来,水利主要是围绕着农业转,传统水利向现代水利、可持续发展水利转变,主要体现在由农业水利向全社会大水利的转变。但是随着经济的发展,水资源供需矛盾日益尖锐,特别是长期以来由于水价关系未理顺,水资源浪费现象极为严重。由于现行水价标准低,水费计收情况较差,没有把水费纳入市场经济体制下的价格体系管理,缺乏市场调节机能。特别是对于农村小型水利工程,由于管理比较粗放,吃"大锅水"、"福利水"的现象仍比较普遍,没有形成有偿供水的机制,严重制约了工程的长久发挥效益。因此,推广应用农村水利供水协会模式，对于规范水价管理、建立合理的水价制度、合理使用水资源有着极其重要的意义。按照市场经济规律，运用经济杠杆，因地制宜地进行农村水利工程管理模式改革，是促进水资源优化配置，节约用水，保证工程正常运行的根本，也是发展壮大农村水利基础设施和基础产业的当务之急。因此，采取有效措施，深化农村水利工程管理体制改革，推广应用灌溉模式、供水协会模式，加强测水、量水工作，建立合理的水价形成机制，是促进农村水利事业发展的关键环节，是非常必要的。

水管理的社会经济作用分析

谭海鸥 林洪孝
山东农业大学，泰安

1 前言

中国历来重视对洪涝旱灾的治理和水的管理，并把它作为治国安邦的大事，自古有"善为国者，必先除水害"的治国方略之说。把水资源作为重要的战略资源予以高度重视，强调"水资源可持续利用是我国经济社会发展的战略问题"。并提出"水是人类生存的生命线，是经济发展和社会进步的生命线，是实现可持续发展的重要物质基础[1]。"因而，在重视技术解决水资源问题的同时，采取了一系列更有效的管理办法，全方位、多层次地强化在合理开发、有效利用、节约和保护水资源中的地位和重要作用。

本文在分析中国缺水和水污染发生原因基础上，通过对水管理在社会经济中作用的研究，结合国内外水管理的做法，系统论述了现行水管理理论方法，以及水资源可持续利用的管理战略和措施，并对今后一定时期水管理工作进行了必要的展望。

2 水资源危机发生的必然性与水管理的重要性

在中国近 30 年左右的时间里，就经历了供大于需、供需基本持平到供小于需的水资源危机过程。从当前看，水资源危机不仅没有缓解，还表现出更加严峻的趋势。

水资源危机的产生，虽与水资源量及时空分布较之经济社会的发展要求不相平衡有关，更与人们开发利用和管理水的指导思想、政策和措施有必然的因果联系。在中国引发水资源危机的原因和类型可归纳为[2-5]：①资源匮乏型缺水。②水污染型缺水。③缺乏合格供水水质型缺水。④供水工程不足型缺水。⑤浪费型缺水。

综上所述，发生缺水危机的核心问题是在经济社会建设与发展中，忽视水资源条件的限制性作用，忽视对水的有效管理，忽视水资源—经济—社会—环境的协调，忽视水源—供水—用水—排水及水处理与回用间的系统性、循环性和相互激励性的关系。由于这些自然和人为的因素，使得经济社会发展到一定程度时发生水资源危机成为必然现象，并显示出加强水管理的日益重要性。

3 国内外水管理研究现状

3.1 国外研究状况

世界各国都十分重视对水问题的研究。美国的水权制度[2]、澳大利亚的水权交易[3]、以色列的节水管理[4]、荷兰的水务管理体制[5]、法国的水务一体化及水价调整办法[6,7]、俄罗斯的水权及水价管理[8,9]等,取得了许多可贵的成效,他们建立在市场化经济体制上的水权、水市场、水务投资融资方式和管理制度等,都对我国水管理改革有重要的参考作用。归纳起来,国外水管理的主要经验如下:

(1)强调水资源统一管理。各国普遍通过立法来保障水资源的公有属性,实行水资源产权的国家所有制度。如美国、法国、英国、澳大利亚等传统实行河岸权的国家,都已在20世纪60年代颁布了水的国家所有制的法律[6]。任何取水行为都必须经过严格的申请、审批和监督管理程序,才能获得水权,即普遍实行水权登记和取水许可制度,获得有限制的取水权是国家或一定区域管理水资源活动,促使有效利用和保护的重要象征,并常常成为管理水资源、配置水资源的重要管理办法。

各国同样通过法律确立代表国家实施水资源统一管理的国家级机构,以及管理一定区域水资源的机构,反映国家水政策和公共意志,以保障水资源的统一管理与配置,妥善协调有关涉水事务及矛盾冲突,保障开发利用水资源的效率和安全。

在强调水资源权属性管理的同时,各国为统一管理、优化配置水资源,广泛开展水资源勘测、调查评价及各类规划的编制等工作,并纳入政府的工作范畴,有的国家还将其用法律固定下来,作为相应管理机构的职责,如美国的《水资源规划法》。

(2)实行流域管理与行政区域管理相结合的管理体制。建立以流域管理为单元,流域管理与行政区域管理相结合的管理体制是各国的基本共同趋向[6]。例如:法国建立的从中央、流域、大区、省级及市镇的水资源、水务统一管理体制;澳大利亚建立的联邦政府对于跨行政区域的河流,实行以流域为单元的水资源、水务统一管理体制。

(3)十分注重城市污水处理和保护水资源环境。治理和保护以城市为重点的水资源环境受到各国的高度重视,并作为城市水务管理的重要工作内容,各国相继颁布相关的法律和严格的管理规范、标准,多实行"谁污染谁治理、谁破坏谁赔偿"的基本管理制度。

欧盟在1970年就开始制定了保护水源和河川的政策,到1990年,欧盟已开始对一般的水源进行管理,并通过了两项立法:一是严格规范市区及郊区废水处理;二是严格规范农业硝酸钠的使用。目前欧盟正在进行解决水源和河川污染的第三次活动,将制定更加严格的制度防止水污染,并将水源保护的范围扩大到更广泛的地面水、地下水及河水和海水等所有的水源和涉及生物化学的使用层面上。

(4)水务市场化成为发展方向。在许多国家,城市水务管理的市场化程度比较高,这不仅仅反映在美国、德国、以色列、澳大利亚等国的水资源产权的流转方面,供水、用水、排水及水处理和回用等多运用市场机制进行管理与运作。对从事水务方面业务的业主多实行委托经营、特许经营等管理运营模式,对用户实行"使用者付费"、"污染者受罚"、"谁污染谁治理"等基本管理制度。

（5）运用价值规律促进水的优化配置与节约保护。在城市水系统的各个环节都能以一定的价格反映其价值，发挥价格在资源配置中的杠杆作用，促进水有效率、效益的运转和管理。例如：用水资源价格（有的称源水价格）、供水价格、排水价格、回用水价格等。把用水的内部成本和外部成本尽量反映出来，如法国里昂市、尼斯市的水价调整管理办法[6]。通过价格手段促进供需水的平衡。

发达国家在水管理方面有许多经验值得借鉴学习，尤其是依法治水，以先进的管理和技术促进节水和水资源环境保护，都对提升我国水管理与运作能力起到很大的促进作用。

3.2 国内研究状况

水管理已成为研究的热点课题，诸多专家学者均撰文探讨我国的水管理体制和可持续发展水资源战略问题[10-21]，但多限于对水源和供水的管理研究，缺乏对水务系统特性、特征和存在的客观规律性及变化趋势，以及与民法相结合的水权；缺乏适合市场经济体制和与《水法》《取水许可制度实施办法》等水法律法规体系相衔接的水权；缺乏水市场、水价、促进节水和再生水利用的水管理体制和运作机制的系统理论研究成果。

4 中国水管理现状及成效

面对复杂的水资源及管理问题，近年来，从治水思路、立法、管理体制和管理办法等诸多方面，积极探索符合中国自然条件和经济社会发展的水管理道路。

4.1 水问题引起了全社会的高度重视，基本树立起了水资源短缺的忧患意识

高度重视水资源的开发利用与保护管理，《水法》[22]以法的形式确立了对水资源应当采取有效的保护管理和提高开发利用效率，把节约用水放到十分重要的地位。

4.2 坚持人与自然和谐共处，不断促进水资源的优化配置

在以水资源的可持续利用支持经济社会的可持续发展新时期治水思路的指导下，水管理坚持人与自然和谐共处思想，在重视生活、生产用水的同时，更加注重生态用水，加强了流域水资源的统一调度，促进了水资源的优化配置，增强了水资源可持续利用的能力。

2001年黄河流域属大旱之年，通过实施水资源的统一调度，在花园口站来水量比多年平均偏少62%的情况下，实现全年不断流。

塔里木河从博斯腾湖向下游成功输水后，结束大西海子水库至台特马湖300 km以上河道近30年断流无水的历史，台特马湖已经形成10 km^2的水面，地下水埋深从过去的13 m上升到平均7.8 m，胡杨长新绿，红柳发新芽，塔里木河下游的生态明显改善。

山东济南市在2001年结束水资源多头管理局面，采取统一调度、依法取水，尤其是加强地下水管理，限采地下水及回灌地下水，以及节约用水等多种手段优化配置水资源，地下水明显回升，在2001年9月17日，沉寂长达926天之久的济南趵突泉喷涌，市区内100多处泉水重涌珠玉，社会反响很好，实现环境效益和经济效益双丰收。

江苏省对苏锡常地区地下水实施封井、禁采措施，1/3的地下水监测井水位开始回升。
太湖流域实施"引江济太"调水，引水后达到或好于III类水质标准（GBZB1—1999）

的水体所占的比例比引水前上升，太湖换水率提高了15%，加快了流域河网及太湖的水体流动，并为向下游供水提供了保证。

4.3 城市水务管理取得较大进展

从城市水务统一管理体制的设置和许多城市水务管理体制改革取得的成效看，主要体现在如下几个方面：

（1）有利于克服城市水务管理体制分割管理的弊端。城市水务统一管理有利于克服部门职能交叉、政出多门、办事效率低下的弊端，体现了机构设置"精简、统一、效能和一事一部"的原则，也有利于部门分工负责、公众参与涉水事务建设与管理。

（2）有利于优化配置水资源。水务部门统筹调度地表水与地下水，优化配置城区、郊区以及区外水资源，能有效缓解供需矛盾，提高城市防洪和供水保证率，改善城市生态环境质量。例如，包头市水务局在黄河水质超标、供水紧张的情况下，实施当地地表水、地下水和黄河水"三水"联合调度，减少黄河取水量，加大水库和地下水的取水量，使供水水质明显改善，保证了百万人口饮水的水量和水质。

（3）有利于促进水资源供需平衡。科学调度整个产业结构和布局，实现水资源的供需平衡。例如，黑龙江省针对呼兰河、穆棱河等流域水田过于集中，影响城镇及农村用水的状况，制定了山东省水田调整战略规划，将水田发展布局向水源丰富的黑龙江、乌苏里江沿岸转移，以减轻中部地区水资源压力，为实现经济社会的可持续发展创造条件。

（4）有利于城市生态系统建设和水环境改善。城乡水资源、地表水、地下水资源统一管理，统筹考虑生产、生活和生态用水，许多城区水环境得到改善，水生态系统建设得到加强，提高了城市品位，拉动了许多城市的地产升值，有利于城市的产业结构调整和现代化建设，如山东滨州市的"四环五海"城市水利建设。同时，使城市防洪工程与市政工程、生态环境建设工程有力地协调结合起来，发挥了综合效益。

（5）有利于建设统一的水务市场和强化水务资产运营，实现水务国有资产的保值增值。例如：上海水务局成立后，组建了水务资产运营管理公司，推进水务工程投资、建设、管理、运营四分离，培育和规范水务工程建设和水务设施运营两个市场。2001年成立了上海水务集团，通过社会投资项目招商，法国通用水务公司购买了原自来水浦东有限公司50%的股权，诞生了我国第一个集制水、输配、销售于一体的中外合资自来水有限公司。

实施城市水务统一管理是水资源管理工作的一次质的飞跃，是资源管理体制的重大改革和创新，对促进可持续发展战略的实施具有特别重要的意义。

4.4 水权理论有力地指导了水管理实践

在社会主义市场经济条件下，水管理迫切需要新的治水理论做指导，水权制度和水市场理论的提出有力地指导了水管理实践。水权与水市场、水资源承载能力与水资源优化配置、水环境保护利用等有密切联系，其实质都是希望从理论上解决如何通过制定和完善相关的政策法规，形成一整套体制、制度、机制来调整水资源配置（包括开源、利用、节流、保护）中的经济利益关系，达到有效管理和提高利用效率的目的。

在水权、水市场理论的指导下，各地进行了积极的探索，漳河上游通过有偿调水化解水事纠纷，水利部在黑河流域建设节水型社会的试点、黄河水量配置等，都取得了显著的成效。

4.5 提出水资源、环境承载能力问题，明确水资源是保障经济社会发展的要素

提出水资源承载能力概念，就对一定的流域或区域开发利用水资源的潜力提出了限制性要求，可将其视为一个"阈值"。它是合理开发利用水资源的底线标准，在该值以上，就可以保障水资源可持续开发利用，支撑经济社会可持续发展；在该值以下，对水的开发利用就会产生诸多恶果，难以支撑经济社会可持续发展。

为此，我国提出并采取了一系列前瞻性的对策和措施。如：节约和保护水资源是国家的一项重大国策；加强水资源的规划管理，合理配置水资源，协调生活、生产和生态用水；大力推行节约用水措施，发展节水型农业、工业和服务行业，建设节水型社会等。

水环境承载能力取决于一定水域在水环境系统功能可持续正常发挥作用的前提下，接纳污染物的能力（纳污能力）和承受对其基本要素改变的能力（系统调节能力）。水环境承载能力是水体纳污的"阈值"。若水体纳污超过了这个"阈值"，将导致水环境结构的破坏并引发某些功能丧失，在其内可保持水环境持续利用，其值大小受生产力水平和人们认知能力的影响，并随物质、文化生活水平的提高会提出越来越高的要求，并将其值逐步向水体自净能力标准靠近，以消除人类活动对水环境可能造成的潜在危害。

水资源承载能力与水环境承载能力是紧密联系、相辅相成的。水管理必须重视对它们的评价，并用其指导水的管理工作。

4.6 需水层次管理理论得到有力的实施

在"以供定需"、"以水定发展"的水资源优化配置要求下，根据经济社会发展对水资源的要求，特别是从人民群众的根本利益出发，针对我国水资源环境日趋恶化的现实，提出了饮水安全、防洪安全、粮食生产安全、经济发展用水、生态环境用水等五个需求层次的理论，并将这一理论用来指导水的管理实践。

满足以上五个层次对水的需求，都是艰巨的任务。目前越高层次的需求，满足程度越低。根据需水层次理论，在水管理中，不仅要尽快地满足人民群众的基本需水要求，而且应遵循可持续发展的需求。提出首先保证基本生活用水和合理安排生产用水，同时，重视生态环境用水。实践中，对生态环境建设采取了积极行动，如加强黄河水统一管理；实施黑河分水方案，应急向塔里木河下游输水；嫩江向扎龙湿地输水，恢复湿地生态；"引江济太"到太湖流域，改善太湖水质等，都使水管理取得了新的成效。

4.7 节水管理促进了节水型社会的建设

在水资源优化配置，计划取水、用水，督促工业、农业及生活用水采用节水型的工艺和设施的基础上，通过调整水价促进节水，促进水管理各项改革措施的落实，使节水管理工作得到实质性的进展，建设节水型社会的目标得到有力实施。例如，农业节水方面，"十五"期末，在全国建设 300 个节水增产重点县，209 个高标准节水增效示范区，并对 99 个大型灌区及 40 个中型灌区进行节水灌溉为中心的续建配套和更新改造，建设了一批国家级节水示范区，新增节水能力近 $11 \times 10^8 \text{m}^3/\text{a}$，第二批 300 个节水增产重点县建设正式启动；工业节水方面，目前全国工业用水重复利用率普遍比 20 世纪 80 年代初提高了 40%以

上。一般工业万元产值新水量从 1980 年的 933 m³（含火电）下降到 90 m³，北京、上海、天津、广州、沈阳、深圳、青岛等市都对工业用水进行定额管理等措施，一些重要城市，如北京、天津等还出台了严格的定额管理、超计划用水加价收费等办法。

通过以节水为中心的提高水资源利用效率工作，在水资源充裕和紧缺地区建设不同的经济结构，量水而行，以水定发展进行经济结构、产业结构调整。推行清洁生产，污水处理和再利用，在用水和退水各个环节采取有效措施的治污工程和非工程措施，全面推动节水型社会建设的步伐。

4.8 以水利信息化为基础和标志的现代化建设，提升了水管理水平

水利信息化是水利现代化建设的基础和重要标志，建成水利信息化是水利实现现代化的战略任务。通过国家防汛抗旱指挥系统、全国水质信息系统和水资源实时监控管理系统等一系列应用系统的建设，提高信息采集、传输的时效性、准确性和自动化水平，为实现水资源优化配置提供手段，为防汛抗旱决策提供依据，为更好地服务经济社会发展创造条件。目前全国水利系统在防汛抗旱、水资源优化调度、水利建设和管理、办公自动化等方面较为广泛地采用了先进的信息技术，信息化建设步伐正在加快。如山东省在山东省水资源费征收管理办法出台后，及时开通了山东省地市、县、用水户与水资源管理部门、财政、银行的计算机联网信息网络，较好地保障了及时掌握征费信息。

20 世纪 90 年代以来的水管理，无论是在管理理论技术、政策法规、体制及运作机制，还是在管理实践方面，都取得了长足进展，提高了人们对水资源的忧患意识，普遍树立了节水观念，促进了水资源开发利用能力和利用效率的提高，为水资源持续利用，保障经济社会持续发展奠定了基础，创造了条件。

5 中国水管理现存问题

中国在可持续发展水管理上，无论是在管理体制、法规建设与执法状况，还是在管理制度、运作机制，以及管理理论的创新等方面，还存在许多亟待解决的问题。

5.1 管理体制性障碍是影响水资源可持续利用的重要因素

传统的水管理体制将城市与农村、地表水与地下水、水量与水质等进行分割管理，严重地违背了水资源的自然循环规律和整体性特征。特别是由此产生的"多龙管水"、"政出多门"，不同水管理机构的职能相互交叉重叠，职责不清，互不协调等问题还比较突出。如除水行政主管部门外，渔业、交通、城建、环保和旅游等部门，都从各自管理的目标出发介入水的管理，使得"开发利用"和"水行政管理"相混淆，水行政主管部门难以代表国家发挥主管部门的职能作用。

5.2 水法律法规有待健全，执法不严的问题仍很突出

市场经济是法制经济，必须建立、健全和完善水法制体系。但在涉水事务的执法管理工作中有法不依、执法不严、违法不究，以及为谋求部门和地区利益而违反法律法规的现象仍然比较突出，主要表现在：①一些地方和部门的领导干部法规观念淡薄，随意干预依

法行政，造成执法困难，这在查处违章取水、拒缴水资源费、审核取水计划等水资源管理活动中比较常见。②一些群众守法意识不够强，破坏水利设施的恶性案件还时有发生。③个别水管理部门遇到水事违法案件，采取消极态度，这在严格执行水资源计划开采、按时足额征收水资源费、处罚水污染案件等方面还比较常见。④不少地方忽视生态环境建设，植树造林力度不够，以致水土流失、水患危害加剧。⑤对水法律法规宣传不力，注重形式，而不注重实效。所以，立法重要，执法更重要，否则难以发挥依法管水的作用。

5.3 水管理对新理论、新技术的发展提出了更高要求

由于我国严峻的水资源环境局势，水管理面临的任务与日俱新，层出不穷，需要尽快解决的重大问题很多。但是，涉水事务管理又是一个新事物，对水管理改革的理论与实践探索更是近年来才展开的工作，在中国提出和加强管理的时期不长，尤其是对水资源规律和水资源开发利用后果的认识，以及采取何种形式管理等，多处于感性认识的阶段，现成的理论、技术方法还不能满足社会实践的要求。例一关于水环境承载能力的理论体系尚未形成。生态问题的提出对水资源合理配置的理论方法形成了新的挑战；例二明晰水权和水权管理理论问题[23]；例三水价构成中的资源水价，环境水价即水资源费、排污权费的征费标准与管理使用，虽然国内外都有一些研究，但远没有达到理论依据充分和能够定量指导实践定价的要求，目前仍主要采用行政定价的办法。

当前在涉水事务管理中要求的新理论、新技术多而广泛，需要相关学科协作攻关，以达到用先进的理论和技术指导和促进水管理工作的发展要求。

5.4 水价偏低阻碍了水务建设与发展

水价应按照价值规律合理调整，使得价格与价值相符合，水利才能形成良性循环的运作机制。水价形成机制不合理，现行水价过低是造成水资源短缺和浪费的主要原因之一。据统计，我国现行水利工程供水平均水价仅为供水成本的 57.5%（张金慧，2001），或者 60%（王丙乾，1997），全国水利工程综合平均水价仅 0.027 元/m^3（张金慧，2001），据国家计委价格司、水利部经济调节司联合调研组在全国"百家大中型水管单位水价调研报告"中指出，2001 年平均生活水价为 0.2395 元/m^3，工业水价 0.2284 元/m^3，农业水价 0.0361 元/m^3。这种水价偏低带来的后果主要有[24]：①水资源难以实现优化配置，供求失衡。②造成水资源浪费、节水工作缺乏动力。③偏低的水资源费难以达到促进节约、保护水资源的作用。④治理工作难以开展。⑤造成供水企业亏损。⑥增加财政负担。所以，水价偏低，是当前影响和阻碍水管理工作改革发展的重要问题，若解决不好，必将削弱水管理工作绩效，难以落实国家相关管理政策、措施。

5.5 城市水务管理体制问题

在城市水务管理体制改革中以下问题还表现得比较严重：

（1）城市区域与流域的协调问题。在强调以城市区域水务一体化管理中，如何协调与流域、与区域之间的关系，克服出现区域割据式的水管理体制弊端。

（2）垄断问题。在水务一体化管理中如何避免一体化管理可能出现的垄断问题，以及因经营管理以利益最大化为目标，使资源管理服务于经营，产生新的水行业壁垒和行业利

益模式,妨碍水行业管理改革的市场化进程,因而,如何将资源管理与用水管理两项职能既分离,又纳入一体化的管理模式中来运作,充分发挥政府和市场的管理与调节作用,应给予足够的重视和研究。

(3) 涉水部门法规的协调与统一问题。由于水务工作涉及的部门、行业较多,覆盖面广,要发挥水资源的综合效益,提高用水效率,就必然要求诸多涉水管理部门、涉水职能交叉部门,就水务问题建立民主协商决策机制,在《水法》法理精神下,在决策层通过充分协商,建设符合水务管理系统要求的政策、法规,这种机制如何建立与运作应深入研究。

(4) 水务管理模式问题。水务管理机构与水务市场的关系,水务管理机构及水务市场模式、管理体制、管理机构设置、管理制度、管理机制等问题。

(5) 水务管理法规建设问题。水务管理法规、规范、标准及执行,应如何适应管理体制改革的进程,并有一定的前瞻性,以为水务管理体制和运作机制改革提供保障,在这方面无论是现行的取水许可制度实施办法、城市节约用水管理规定、各地的水资源管理条例等都存在与国家要求的水务管理方针、政策不相协调的内容,如何修改完善,建立符合实际情况的水资源及城市水务管理法规体系是近期应加快解决的。

城市水务管理体制改革的诸多方面适应了逐步建立和完善社会主义市场经济体制的要求,适应了经济全球化的时代发展潮流,顺应了经济社会发展对水务管理提出的改革创新要求,但一定要将各地的实践经验认真总结提炼,并上升到理论高度,以进一步指导城市水务管理体制改革。

6 关于水资源可持续开发利用与管理的思考

发展是人类社会永恒的主题。如何处理人与水、经济与水、社会与水及发展与水的关系,不仅是技术问题,更是社会和伦理问题。认清产生问题的原因是必要的,采取辩证积极的态度处理问题,克服发展中的困难和与水资源环境的冲突,才可达到人类社会与自然的和谐,才可实现人类社会的可持续发展。

6.1 对水资源环境与人类关系的认识[24, 25]

在人与水的关系上,应重视水资源及其相互依存的自然环境的客观容纳能力,运移变化规律,注重协调好人、经济、社会与资源、环境间的系统性、整体性及有机性的关系。发生缺水时,应多从供、用水行为中寻找原因,寻求解决问题的办法,维护资源环境系统与人类社会系统的和谐。总结出一条人与自然、人与社会、人自身行为间和谐地相依相伴、相生相长的资源环境可持续开发利用与经济社会可持续发展的道路。

6.2 思辨关系的转变

革除发展中的"资源高消耗、环境高污染、人类高消费"的模式。在人与自然的关系上,要适当立法,立法中应遵循天道原则、人道原则、自然与人类和谐的原则。人的一切行为都要维护和促进自然与人类关系的和谐[26]。

6.3 解决水资源危机是人类社会的紧迫任务

缺水势必严重影响粮食生产、工业发展、人民生活和社会稳定，以及环境卫生状况。国民经济建设与发展每上一个台阶都必须有适量安全的用水作为支撑，否则，不仅不能实现发展，还会影响和破坏原有已经取得的成效，"不进则退，不用则废"的基本规则，同样适用于水对经济社会建设与发展的保障作用。

到 2030 年我国人口将达到 16 亿，人均占有水资源量会减少 1/5。在今后几十年里我国经济增长仍将处于快速发展期，到 21 世纪中叶，国民生产总值要增长 10 倍以上，城市和工业需水必将大幅度增加，同样污废水排放量也将增加。若不能采取十分有效的措施解决水问题，人水冲突、经济社会发展与水的冲突、生态环境恶化与水的冲突，都将会更加严重。同时，还应看到，在这些冲突矛盾中，人、社会、环境建设总是处于次要的地位、被动的地位，人、社会、环境离不开水，水可以离开他们。人类历史经验证明，在处理与水的关系问题上，运用"征服"和"战胜"总是不恰当的，事后的"惩罚"常是根本性的破坏，会使多年的建设成就毁于一旦，并使人类陷入更难实现发展的泥潭。水污染、用水浪费、计划经济体制下建立起来的僵化的水管理体制等问题，都与人总把自己放在矛盾冲突的主要地位有关，而难以达到与水与自然的和谐境界。

6.4 实施可持续发展战略是解决水资源危机的有效途径

当今人类面临的人口增长、资源消耗、环境污染、粮食生产和工业发展等全球性问题，罗马俱乐部（The Club of Rome, 1968）发表了题为《增长的极限》的报告，认为地球的支撑力将会在下个世纪（指 21 世纪）某个时段内达到极限，经济增长将发生不可控制的衰退，其避免和解决的最好方法是限制增长，即"零增长"。爱德格尔从哲学的角度认为现代技术是对人的本质的异化，技术将人与自然的多样性联系给抹杀了，技术将自然变成单一的、齐一的、功能化的物质，使得人远离自然，使人自己丧失了人本质的多元性。马尔库塞从社会科学史角度认为现代技术已成为科学的形式，使西方陷入"单向度"社会和"单向度"人。罗蒂从文化角度认为科学是后神学时期的产物，自然科学取代宗教而成为当代人的文化中心。上述现代西方科技思潮对人在自然系统中作为的评价，虽有些悲观和偏激，但也不能不引起人类的反思，养育人类的自然为什么成了人类发展的限制性因素，人应如何辩证地看待自己的发展，选择自己的发展，应如何辩证积极地处理与自然的关系，都是应该深入思考的。

对解决全球性水资源危机，既要认识到问题的严重性，也应保持谨慎乐观的态度。正如甘哈曼针对罗马俱乐部的悲观观点认为，要认识历史，要预测未来的历史，应采取历史外推的方法，人类的资源在人自己，没有极限的增长、资源的无限。正如马克思、恩格斯指出，每一次划时代的科学发现，都改变了人们认识和改造世界的世界观和方法论。人类与水的关系的认识及处理的科学技术方法，概不例外，实施水资源可持续利用战略是解决水资源危机的有效道路。

在水资源开发利用与管理中应该做到[26]：①水资源开发利用、环境保护和经济增长、社会发展协调一致。②水资源及其依存的自然生态系统，对国民经济建设和社会发展的承载能力是维持水资源供需平衡的基础，是制定水资源开发利用战略措施的出发点和着眼点。③水资源的开发利用措施和管理办法,必须保障自然生态系统的良性循环和发展。④必须运用系统科学的方法研究水资源—经济—社会—环境复合巨系统,并用动态的、辩证的观点研究这个开放复合巨系统的变化规律。

7 结语

本文在分析发生水资源危机的必然性原因和加强水管理重要性的基础上,研究了国内外水管理的现状及发展趋势,分析了中国水管理取得的主要成就,以及存在的一些问题,提出了水资源可持续开发利用及管理保护应采取的思想、应遵循的原则,以及理论方法,以有助于强化水的管理,强化水的开发利用成效。上述想法供同仁们参考。

参考文献

[1] 汪恕诚. 水权管理与节水社会[J]. 中国水利, 2001 (5).
[2] John R. Teerink, Masshiro Nakashima. 美国、日本、水权、水价、水分配[M]. 刘斌, 等译. 天津: 天津科学技术出版社, 2002.
[3] 孟志敏. 国外水权交易市场[J]. 水利规划设计, 2001 (1).
[4] 汪光焘. 城市节水技术与管理[M]. 北京: 中国建筑工业出版社, 1994.
[5] 成建国. 水资源规划与水政水务管理[M]. 北京: 中国环境科学出版社, 2000.
[6] 林洪孝. 水资源管理理论与实践[M]. 北京: 中国水利水电出版社, 2003.
[7] Smith V K. Resources valuation at a crossroad. Estimating Economic Values for Natur: Methods for Non—market Valuation, 1996.
[8] Fakhraet, SH. And R. Naraganan, Price Rigidity and Quantity Rationing Rules Under Stochastic Water Supply Water Resource Res., 1984, 20 (6).
[9] 岳梦熊. 坚持开源节流并举,努力建设节水型社会[J]. 中国水利, 2001 (4).
[10] 石玉波. 健全体制 创新机制 推进水务工作改革与发展[J]. 中国水利, 2002 (4): 14-16.
[11] 张岳. 流域管理与行政区域管理相结合的新体制势在必行[J]. 中国水利, 2002 (8).
[12] 吴季松. 水务管理体制改革的发展方向[J]. 中国水利, 2002 (7): 35-39.
[13] 汪恕诚. 资源水利的本质特征、理论基础和体制保障[J]. 中国水利, 2002 (11): 6-8.
[14] 钟玉秀, 等. 合理的水价形成机制初探[J]. 水利发展研究, 2001 (2): 16-16.
[15] 吴季松. 从海牙会议看国际水资源政策的动向和潮流[J]. 中国水利, 2000 (7).
[16] 任光照, 等. 城市用水[M]. 北京: 中国水利水电出版社, 1989.
[17] 翟浩辉. 关于水利现代化问题[J]. 中国水利, 2002 (7): 8-15.
[18] 沈大军, 梁瑞驹, 王浩, 等. 水价理论与实践[M]. 北京: 科学出版社, 1999.
[19] 甘泓, 王浩, 等. 水资源需求管理[J]. 中国水利, 2002 (4): 66-68.
[20] 姜文来. 水资源价值论[M]. 北京: 科学出版社, 1999.
[21] 李曦, 熊向阳, 等. 我国现代水权制度建立的体制障碍分析与改革构想[J]. 水利发展研究, 2002 (3).
[22] 中华人民共和国水法[M]. 北京: 中国水利水电出版社, 2002.

[23] 汪恕诚. 努力推进水资源可持续利用 为全面建设小康社会作出贡献[J]. 中国水利，2003（1A）：8-18.

[24] 林洪孝. 城市水务系统管理模式及运作机制研究[P]. 西南交通大学，2003（10）：20-22.

[25] 林洪孝，赵强. 关于人类社会与水资源可持续利用关系的探索[J]. 水利发展研究，2004（2）：30-32.

[26] 林洪孝，等. 关于水资源可持续开发利用的思考[J]. 水土保持学报，1999（10）：41-43.

水资源陆海空协同系统研究[*]

赵然杭
山东大学土建与水利学院，济南

我国是一个水资源短缺、时空分布不均的发展中国家。进入 21 世纪，我国又连续遭遇严重干旱，旱灾发生的频率与影响范围扩大，持续时间与干旱造成的经济损失日益增加。由于社会经济的迅速发展、人口不断增加，水资源短缺已经成为我国实现建设小康社会目标的重大制约因素。估计到 2030 年我国人均水资源占有量将从现在 $2200\,\text{m}^3$ 减少到 $1800\,\text{m}^3$ 左右，需水量接近我国河川流域地表与地下水资源可开发利用的总量，干旱缺水问题将更加突出[1]。因此，增加我国可开发利用的水资源总量，应是 21 世纪我国水资源开发利用中的一个具有战略意义的重要命题。

1 陆海空水资源系统协同方程

陆地（地表与地下）水资源系统的水量平衡微分方程式为[2]

$$dW/dt = F(t) - O(t) \tag{1}$$

式中，W——系统在时间 t 的蓄水量；$F(t)$——时间 t 河川（地表与地下）输入系统的水量；$O(t)$——时间 t 系统的总用水量，包括蒸发、渗漏等损失水量。在河川径流尚未得到开发的自然条件下，显然

$$O(t) \leqslant F(t) \tag{2}$$

由于地表径流的随机性，$F(t)$ 呈现随机多变的特点，在自然条件下完全不能适应系统用水量及其过程变化的需求，因此在流域干、支流上修建蓄、引水等各种控制性水利工程对自然水文系统进行开发，形成陆地水资源系统。

文献[3]依据协同学原理[4]导出的水资源系统可持续发展状态方程为

$$dq/dt = N(q,a) + F(t) \tag{3}$$

$$dq/dt = N(q,a_0) + F(t) \tag{4}$$

式中，q——陆地水资源系统可持续发展状态矢量；a——从外流域调入的水量；a_0——

---

[*] 基金项目：水利部科技创新项目 scxc2005-01；山东省水利厅项目。

海水淡化水资源量；$N(q,a), N(q,a_0)$ 均为非线性函数。

其中 $N(q,a) = a + f(q,a)$，$N(q,a_0) = a_0 + f(q,a_0)$；$f(q,a) = O(t)$，$f(q,a_0) = O(t)$。

由于状态 q 等价于陆地水资源系统的蓄水量 W，则式（4）可表示为

$$dW/dt = N(W, a_0) + F(t) \qquad (5)$$

式中，$N(W, a_0) = a_0 - f(W, a_0)$；$f(W, a_0) = O(t)$。

式（5）是陆地与海水淡化水资源系统的协同方程。

以式（5）为基础，文献[2]考虑人工增雨，使系统输入水量 $F(t)$ 中增加一部分云水资源转化雨量 $\Delta F(t)$，则式（5）变为

$$dW/dt = N(W, a_0) + F(t) + \Delta F(t) \qquad (6)$$

式（6）是陆海空水资源系统协同方程。由此可见，从增加我国水资源可利用总量的目标出发，建设传统陆地水资源与非传统水资源——海水、云水的水资源陆海空协同系统，可使目前单一陆地水资源系统向着陆地水资源、海水资源、云水资源的协同系统发展。

2 陆海空水资源系统的协同原理

水资源可持续发展系统结构功能的形成是一个历史演变过程。水资源系统具有自然与社会双重属性，水资源的自然特性远在人类出现以前就已存在。但是，近几十年来，由于社会经济的迅速发展和人口的剧烈增加、水资源供需紧张和生态环境的急剧恶化，人们逐渐认识到，衡量水资源开发与管理的效应，不仅仅是经济效益，更是社会、经济和生态环境效益的综合。随着人类社会的发展，水资源系统的规模、结构和功能更加复杂化和综合化，应该向着社会经济和生态环境良性循环的可持续方向发展。

从系统科学的观点看，水资源系统可持续发展的实现，就是水资源—生态环境—社会经济系统可持续发展结构功能不断完善的体现。只有水资源系统中生态环境、社会经济结构合理，才能取得功能的整体最优；只有系统有序稳定地演变才能取得系统可持续发展的完善实现。因此，以陆地水资源为核心，充分利用海水和云水资源，使陆海空水资源系统协同工作，是水资源系统可持续发展的关键。它不但使水资源系统结构功能更加合理，能够提高系统供水的可靠性、稳定性和抗干扰能力，向着有序的良性循环方向发展，而且也符合复杂水资源系统具有自然与社会属性的实际情况。

2.1 河海水资源的协同

根据式（6）可见，在我国沿海缺水城市，如果在水资源可持续发展系统中开发水源—海水淡化，设地区年海水淡化水资源量 a_0，当河海水资源系统联结后，则河海水资源协同系统的水量平衡微分方程为

$$dW/dt = F(t) - O(t) + a_0 \qquad (7)$$

取时间 t 为 $1\,a$，$f(W, a_0) = O(t)$，则式（7）可转换为式（5）。

由此可见，河与海水资源系统协同工作，则系统可持续发展状态与用水安全性会有明

显改善,主要反映在两个方面:①海水淡化量的非随机高度稳定性,可以提高系统供水的可靠性与稳定性;②削弱河川径流随机性对系统的影响,即提高了系统抗干扰能力。由于经济技术等原因,取之不尽的海水没有得到充分利用,随着高新科技的发展,海水淡化成本的不断降低,海水作为一种稳定水源,具有独特的优势和良好的应用前景[3]。

2.2 陆空水资源的协同

地球上的水资源可分为大气水、地表水和地下水,云水资源是针对大气水而言。面对淡水资源的日益匮乏,节流与开源成为解决水资源短缺的有效途径。节流包括节约用水、合理用水和水资源的综合利用;开源则需要在现有可利用水资源之外,寻找新的淡水来源。由于地球上的淡水(地表水和地下水)都来源于自然降水,因此,开发利用雨、云水资源是解决水资源短缺的根本途径。云水资源开发后,如果不合理利用,则与普通降水一样,多数汇入地表径流流失,仍然不能进入土壤和作物体内,这样就会失去开发的意义,因此一定要把开源与节流结合起来,即将云水资源与陆地水资源协同开发利用。

2.2.1 空中云水资源

人工增雨是开采空中云水资源的有利途径。空中水汽形成降雨必须同时具备两个条件:(1)空气中的实际水汽压要超过同温度空气中的饱和水汽压或者空气的相对湿度;(2)空气中要有足够数量的凝结核[9]。人工增雨就是在一定的云层条件下,通过人工干预,提高降水效率,从而增加自然降水量。有关资料表明,在每次降雨过程中,人工增雨一般能增加20%左右降雨量[11]。因此蕴藏在云中的大气水成为干旱缺水地区宝贵的水资源。

水资源陆空协同系统的建立,可使我们具备对空中水资源的调控能力,加强陆地水资源同空中水资源的联系,统筹规划,在合适的时间,合适的区域,将空中水资源转化成降雨,以增加水资源可利用总量,缓解水资源紧缺的现实。

2.2.2 陆地雨洪水水资源

根据公式(6),陆地(地表与地下)水资源是水资源系统的核心,通过各种水利工程措施,大力开展雨洪水资源利用,是开发陆地水资源的有效途径。雨洪水利用即通过地表微地形改变,以改变雨洪水地表径流汇集方式或改变地表径流运动路径等达到径流局部汇集或增强入渗能力,来实现雨洪水资源的充分利用。对于一个流域或封闭区域,雨洪水利用主要有三种方式。

①微地形改变雨洪水就地利用　这种方式是指通过地表微地形的改变,如夷平、垄起等增加地表土壤入渗能力,或者聚集雨洪水就地利用的一种方式。水土保持措施中的梯田、鱼鳞坑、水平沟、水平阶等就是这种方式的具体应用范例。

②微地形改变雨洪水叠加利用　这种方式是在微地形改变雨洪水就地利用的基础上,将邻近地表雨洪水汇集其上加以利用的一种叠加利用方式。如我们常常见到的隔坡梯田等。

③微地形改变雨洪水异地利用　这种方式是指通过修建或利用已有集流场,将集流场的雨洪水蓄集在修建的蓄水设施中供异地利用的一种方式。一般由集流场、蓄水设施及利用设施三部分组成,或者说由集流系统、蓄水系统及利用系统构成。

由于自然条件和技术经济水平的限制,即使采取各种雨洪水资源化方式,雨洪水资源也不可能完全被开发利用，参考联合国粮农组织（FAO）提出的有效降水量概念，我们将雨洪水资源化可实现潜力定义为：在一定自然和技术经济条件下，通过已有的利用方式和技术，雨洪水资源中可以开发利用的最大量[5]。其一般公式表达为

$$R_a = \lambda_R \times P \times A \times 10^3 \tag{8}$$

式中，R_a——流域或封闭区域雨洪水资源可实现潜力，m³；P——流域或封闭区域降水量，mm；A——流域或封闭区域的面积，km²；λ_R——降雨调控系数，与技术、经济水平、工程措施等有关。

3 实例研究——以济南市为例

济南市是山东省省会，著名的泉城和历史文化名城，但又是一个水资源短缺的城市，水资源供需矛盾十分尖锐。根据济南市属于内陆城市的实际情况，以陆地（地表与地下）水资源为核心，大力开展雨洪水资源利用，并充分利用空中云水资源，建立济南市水资源陆空协同系统。根据公式（6）可得

$$\frac{dW}{dt} = N(W, R'_a) + F(t) + \Delta F(t) \tag{9}$$

式（9）为济南市水资源陆空协同方程。式中 R'_a——雨洪水资源化增加的陆地水资源系统输入水量，$N(W, R'_a) = R'_a - f(W, R'_a)$；$f(W, R'_a) = O(t)$。显然

$$R_a = R'_a + F(t) \tag{10}$$

3.1 济南市雨洪水资源开发利用

济南市地貌区划自南而北分为三区：低山丘陵构造剥蚀区（也称南部山区）；山前平原侵蚀堆积区（主要是城区）；黄河—小清河冲积平原堆积区。根据济南市的地形特点和雨洪水产、汇流情况，考虑济南市的地下水补给，本文重点研究城区和南部山区（研究区域范围为：东至巨野河，西到玉符河，南至分水岭，北到黄河）的雨洪水资源化问题。以此为基础，建立济南市的雨洪水综合集流（截渗）—存储—传输—利用系统。

3.1.1 济南市雨洪水资源化方式

根据济南市城区雨洪水利用的优越条件：具有雨水资源化开发利用的降雨条件；有利的地形；小清河、大明湖可作为雨洪水调节湖；城市公共用地面积大，排水管网完善，雨水收集条件优越；雨水回灌条件好等，结合济南市的实际情况，济南市城区可采取以下雨洪水利用方式。

①微地形改变雨洪水就地下渗[6]。下凹式绿地是最好的渗透设施，为此，宜将城市公园、草坪、苗圃、公路绿化带等现有绿地改造成良好的入渗场地来接纳居民区和道路上的雨水径流；将城区不透水地面改造成透水地面，诸如在人行道上铺设透水方砖、草皮砖、步行道以下设置回填沙石、砾料的渗沟，采用透水性沥青路面、混凝土透水路面等，以增加入渗量；还可以通过在城市适宜地段营造渗透池、渗透孔、渗透槽、渗透管等方法加大对雨水的渗透量。

②微地形改变雨洪水异地利用。城市路面、屋面、庭院、停车场及大型建筑等可作为集水面，通过导流渠道将雨水收集输送到蓄水设施；充分利用城市已建雨水排水管网体系，结合城市地形走向，规划、建立综合性、系统化的蓄水工程设施；当地面土地紧缺时，建设地下蓄水池。将收集的雨洪水通过传输—净化—利用系统，开展雨洪水利用。

③雨水回灌[7]。济南市由于过量开采地下水，导致沉降漏斗范围不断扩大，不少地方甚至出现了严重的地面沉降和断裂带，所以采取有效措施，进行雨水回灌地下水，势在必行。根据济南市区地质、地形、周围环境等条件，可以对现有的两用井、渗井等加以利用。其中渗井是深层入渗工程，它既可补充浅层地下水，也能补充深层地下水，是进行雨水回灌的有效途径。

南部山区是济南泉群的直接或间接补给区[8]。根据济南市南部山区地层分布与地质构造、地层岩性与降雨产流、断层控制岩溶水流向等地质及水文特征，按照三种主要的雨洪水利用方式，重点采取各种拦水下渗补源工程措施（田间、田外、强渗区、河道内、河道与强渗区的联合利用等工程措施），开展雨洪水利用。

田间工程措施包括平整土地、水平梯田等各种水土保持措施，即雨洪水就地利用；田外工程措施主要是雨洪水异地利用，建立收集—储存—传输—利用系统，以供给农业及人畜饮用水；在南部山区24个强渗区及小清河支流上，通过大量水文特性、地质构造、下渗等水文、水力分析和野外勘察及现场实验，在不妨碍行洪的前提下，修建一批大口径渗井或"滚水坝"（橡胶坝等），迫使降雨或上游弃水下渗直接补充各大泉群水量；将玉符河上卧虎山水库的水，经过自流引到市区近郊的强渗区内，通过河流与强渗区的联合调度，开发利用雨洪水资源，下渗补充泉群地下水。

3.1.2 济南市雨洪水资源化可实现潜力分析计算

根据公式（8），雨洪水资源化可实现潜力的分析计算主要是不同利用方式下降雨调控系数 λ_R 的确定。济南市雨洪水资源化可实现潜力 R_a 分为两部分，即城区雨洪水资源化可实现潜力 R_{a_c} 和郊区雨洪水资源化可实现潜力 R_{a_j}。

（1）城区雨洪水资源化可实现潜力 R_{a_c} 的确定。

城区雨洪水资源化可实现潜力 R_{a_c} 的确定，取决于城区日用雨洪水量 Q 与城区调蓄雨洪水的蓄水池总容积 V。确定过程如下：

$$\left.\begin{array}{ll} \left.\begin{array}{ll} R_{a_{ci}} = W_{ci} & W_{ci} < V \\ R_{a_{ci}} = V + T \cdot Q & W_{ci} \geq V \end{array}\right\} & \Delta V \leq 0 \\ \left.\begin{array}{ll} R_{a_{ci}} = W_{ci} - \Delta V & W_{ci} < V \\ R_{a_{ci}} = V + T \cdot Q - \Delta V & W_{ci} \geq V \end{array}\right\} & \Delta V > 0 \end{array}\right\} \quad (11)$$

则，

$$R_{a_c} = \sum_{i=1}^{n} R_{a_{ci}} \quad (12)$$

式中：T——降雨历时（天数），ΔT——前后两次降雨的时间间隔（天数），$\Delta V = V - \Delta T \cdot Q$，$W_{ci}$——第 i 次降雨产生的城区径流总量，$i=1,2,\cdots,n$，n 为降雨次数。

（2）郊区雨洪水资源化可实现潜力 R_{a_j} 的确定。

郊区雨洪水资源化可实现潜力 R_{a_j} 等于郊区雨洪水资源量与郊区雨洪水可利用率 κ 的

乘积，而郊区雨洪水资源量包括郊区降雨径流总量 W_J 与城区雨洪水弃水量（$W_C - R_{a_c}$）之和，用公式表示为

$$R_{a_j} = (W_J + W_C - R_{a_c}) \cdot \kappa \qquad (13)$$

式中：W_C——年历次降雨城区产生的径流总量，10^4m^3；W_J——年历次降雨郊区产生的径流总量，10^4m^3。

若不考虑城区雨洪水弃水量，即当 $W_C - R_{a_c} = 0$ 时，则有

$$R_{a_j} = W_J \cdot k \qquad (14)$$

由公式（14）和公式（8）可知

$$\lambda_R = \alpha' \cdot k \qquad (15)$$

显然，当某一流域或封闭区域降雨径流系数一定时，降雨调控系数 λ_R 的值随着雨洪水可利用率 κ 的变化而变化，而雨洪水可利用率受工程措施和非工程措施的影响很大。

（3）济南市（研究区）雨洪水资源化可实现潜力 R_a 的确定。

济南市（研究区）雨洪水资源化可实现潜力 R_a，即城区雨洪水资源化可实现潜力 R_{a_c} 和郊区雨洪水资源化可实现潜力 R_{a_j} 之和，用公式表示为

$$R_a = R_{a_c} + R_{a_j} \qquad (16)$$

济南市区主要为小清河流域，根据小清河黄台桥以上流域实测降雨径流资料，可计算每次降雨相应的径流总量 W_i[10]。

$$\Delta W_j = 0.36(Q_j + Q_{j+1}) \cdot \Delta t_j \big/ 2 \quad (j = 1, 2, 3, 4, \cdots, m)$$

$$W_i = \sum_{j=1}^{m} \Delta W_j \qquad (17)$$

式中：Q_j、Q_{j+1}——分别为第 j、$j+1$ 时刻的实测流量，m^3/s；ΔW_j——第 j 时段的时段径流量，10^4m^3；Δt_j——Q_j 与 Q_{j+1} 之间的时间间隔，h；m——实测地表径流时段数；W_i——第 i 次降雨的径流总量，10^4m^3。

对于每场降雨，黄台桥以上流域降雨径流总量 W_i 由不透水区域（城区内硬化面积）产生的径流量 W_{i1} 和透水区域（郊区和城区内的非硬化面积）产生的径流量 W_{i2} 两部分组成。不透水区域产生的径流量 W_{i1} 可以用不透水区域本次平均降雨深 P_i 减去一个初损值 ΔP_i（主要考虑填洼等初损及降雨初期水质较差而弃水，其值经过大量实验分析确定）得到净雨深，净雨深与不透水面积 A_1 的乘积即为不透水区域产生的径流总量，可用下式表示。

$$W_{i1} = 0.1(P_i - \Delta P_i) \cdot A_1 \qquad (18)$$

则本次降雨透水区域产生的径流总量为

$$W_{i2} = W_i - W_{i1} \qquad (19)$$

则本次降雨透水区域的径流深 R_2 和相应的径流系数 α'，用公式表示为

$$\left.\begin{array}{l}R_2 = \dfrac{10W_{i2}}{A_2} \\ \alpha' = \dfrac{R_2}{P_i}\end{array}\right\} \quad (20)$$

式中：A_2——透水区域的面积，km^2。

由此可以计算出每场降雨郊区产生的径流量W_{Ji}，公式为

$$W_{Ji} = 0.1 \cdot \alpha' \cdot P_i \cdot A_J \quad (21)$$

式中：A_J——郊区的面积，km^2。

则每场降雨城区产生的径流量W_{Ci}为

$$W_{Ci} = W_i - W_{Ji} \quad (22)$$

由此可得：
$$\left.\begin{array}{l}W_C = \sum_{i=1}^{n} W_{Ci} \\ W_J = \sum_{i=1}^{n} W_{Ji}\end{array}\right\} \quad (23)$$

通过对济南市生态环境用水和部分工业用水量的分析，经过公式（11）—（23）的计算，2003年研究区域雨洪水资源化可实现潜力R_a随城区蓄水池总容积V的变化曲线如图1。

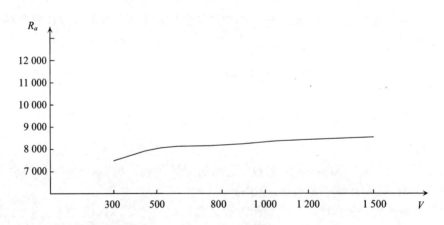

图1 雨洪水资源化潜力R_a与城区蓄水池总容积V关系曲线

注：研究区域$Q=15 \times 10^4 m^3$、$k=50\%$。

3.2 济南市空中云水资源开发利用

济南市属暖温带半湿润季风气候，春季干旱多西南风，夏季炎热多雨，秋季气爽宜人，冬季寒冷多东北风。降雨有明显的季节性，6—9月为汛期，7、8月份占全年降雨量的50%；降雨在空间上分布不均，济南市多年平均降雨量在750～580 mm，自东南向西北递减。根据济南市近15年各级降雨日数资料分析得到，日降雨量≥5.0 mm的年平均降雨日数为28.1天，其中有16.1天集中在夏季。春末至秋初（5—9月）又以积状对流云降雨为主，

多达 65%以上，积状云含水量充沛，自然降水效率较低，降水量仅占云中含水量的 10%～20%，是进行高炮、火箭人工增雨较理想的云层[11]。济南市广大的南部山区 2—9 月年平均降水量为 738.4 mm，根据国内外大量人工增雨试验结果表明，在科学指导下的人工增雨作业，可增加降水量 20%左右，即通过人工增雨，济南市南部山区 2—9 月年平均可增加降水量 147.68 mm，使济南市（研究区）空中云水资源 $\Delta F(t)$ 达到 $13\,646\times10^4\ m^3$。

南部山区石灰岩裂隙溶洞发育，易于大气降水的渗入和地下水的积蓄，是济南市地下水的补给区。因此，根据南部山区的地理特点，并结合雨洪水资源开发利用，在南部山区合理布设作业炮点，运用科学的作业手段，合理地开发南部山区春初至秋末空中云水资源特别是汛期水资源，是完全可行的。

3.3 济南市水资源陆空协同工作

由上述雨洪水资源化潜力关系曲线可知，当 $V=1\,000\times10^4\ m^3$ 时，济南市（研究区）雨洪水可利用量为 $8\,325\times10^4\ m^3$。若考虑空中云水资源的开发利用，经计算可增加雨洪水可利用量 $2\,039\times10^4\ m^3$，由公式（9）和（10）可知，若使济南市水资源陆空协同工作，则济南市（研究区）雨洪水可利用总量达到 $10\,364\times10^4\ m^3$。济南市保泉条件下的水资源供需平衡分析结果显示[6]，2005 年保证率为 95%时济南市区的缺水量为 $11\,295\times10^4\ m^3$，因此，大力开展雨洪水资源开发利用，并充分利用空中云水资源，实施水资源陆空协同工作，不但能够减轻城市防洪压力，而且对缓解济南市水资源短缺问题起重要作用。

4 结语

本文依据陆海空水资源系统协同方程，从增加我国水资源可利用总量的目标出发，论述了以陆地水资源为核心，充分利用空中云水资源，并推广应用海水淡化技术，加大对海水资源的开发利用，使水资源陆海空系统协同工作，从而建立水资源陆海空协同系统，以缓解我国水资源越来越短缺的现实。

根据济南市属于内陆城市的实际情况，重点研究了济南市陆空水资源系统中有关雨洪水资源利用问题。指出雨洪水资源化是济南市陆地水资源开发利用的核心，积极开展人工降雨活动是开采空中云水资源的有效手段。使济南市水资源陆空协同工作，是解决济南市水资源短缺、恢复泉城特色、减轻城市防洪压力、提高水资源利用效率、实现济南市水资源与社会经济可持续发展的主要途径。

参考文献

[1] 汪恕诚. 水环境承载能力分析与调控[J]. 水利规划设计，2002，1：3-7.

[2] 陈守煜. 论 21 世纪我国水资源开发利用的陆海空协同系统[C]//水与社会经济发展——中国水问题论坛第三届学术研讨会论文集. 北京：中国水利水电出版社，2005.

[3] 陈守煜. 21 世纪水资源开发利用的河海协同原理[J]. 大连理工大学学报，2003，6.

[4] 哈肯. 协同学：理论与应用[M]. 北京：中国科学技术出版社，1990.

[5] 吴普特，黄占斌，高建恩，等. 人工汇集雨水利用技术研究[M]. 郑州：黄河水利出版社，2002.

[6] Shizuo Shindo. Rainfall harvesting in Volcanic Island of Japan-Case studies of the Izu Island[A]. Proceeding of International Symposium & 2nd Chinese National Conference on Rainwater Utilization[C]. Xuzhou, Jiangsu Province, China, 1998: 131-133.

[7] S Xu, R Higashijima, et al.. Improvement plan of Y river through downtown of F city, proceedings of the eighth international conference on urban storm drainage, Vol.1, pp.218-225, 1999, Sydney, Australia.

[8] 王琳, 张祖陆. 济南市南部山区生态恢复与重建途径探讨[J]. 地理与地理信息科学, 2003, 5（3）: 71-75.

[9] 何栋材. 开发云水资源, 增强区域防灾减灾能力[J]. 甘肃科技, 2002（6）: 65.

[10] 叶守泽. 水文水利计算[M]. 北京: 水利电力出版社, 1992.

[11] 龚佃利. 山东省主要降水系统云水资源转化特征[J]. 山东气象, 1995, 4.

莱州湾滨海流域农业灌溉用水对社会经济的影响及对策

曹彬[1]　王维平[1]　范明元[2]　黄继文[2]
1. 山东省济南大学资源与环境学院，济南
2. 山东省水利科学研究院，济南

1　农业灌溉用水给莱州湾滨海地区社会经济发展和生态环境带来严重影响

莱州湾海咸水入侵区[1]位于北纬36°25′～37°47′，东经118°17′～120°44′，涉及山东省4个地市8个县市区，包括烟台市（龙口市、莱州市、招远市）、青岛市（平度市）、潍坊市（昌邑区、寒亭区、寿光市）、东营市（广饶县），海咸水入侵面积1000 km²。区域形状像个月牙状，东西长200 km，南北宽40 km，海水入侵面积733.4 km²。从南到北分布着山区、丘陵和平原，山丘区修建了大量的地表水拦蓄工程，平原区以开采地下水为主供水，来满足社会经济的发展。由于长期在山区河流中上游大量拦蓄地表水，减少了入海径流量和地下水补给量，而平原井灌区农业生产快速发展，粮食产量增加很快，特别是自1976年后，机井数量急剧增加，地下水开采量相应也增大。当遭遇到20世纪80年代初期的连续干旱年，地下水采补不平衡，导致严重的海咸水入侵并不断发展，给当地社会经济和生态环境带来巨大的不利影响，特别对农业是致命的打击。过去滨海河流冲洪积平原区，土质肥沃，地下水丰富，农业生产比较发达，粮食亩产达500 kg以上。但是现在海咸水浸染后，地下水变咸，土壤产生不同程度的盐渍化，盐碱地面积不断扩大，良好的耕地生态环境逐步遭受破坏[2]。据不完全统计，全区已有67.2万亩耕地处在浸染区，7446眼机井报废，50万亩耕地丧失了灌溉能力，5万多亩耕地产生了次生盐碱化，农业产量大幅度降低。根据减产成数估计，全区每年减产粮食1.5亿～2亿kg。对工业的影响，海咸水入侵前，本区工业地处海滨，地理位置优越，环境优美，工业用水方便，生产、生活条件一般都比较好，产值利润也高。海咸水浸染后由于地下水被污染，水质变咸，供水井报废，工业供水发生危机，给生产、生活造成了很大困难。海咸水浸染灾害使工业设备严重锈蚀，产品质量下降，经济效益降低，部分企业面临转产或搬迁的困境。据估计，平均每年工业产值损失3亿～4亿元。地下水浸染后，人畜吃水发生了严重困难，据不完全统计，本区有39万人吃水困难。海咸水浸染区，群众不得不长期饮用劣质水，危害人民身体健康。

2　莱州湾海咸水入侵是涉及千千万万农民的农业生产发展、生存环境的社会经济问题

莱州湾海岸带以莱州市虎头崖为界，以西为泥质海岸，以东为砂质海岸。因此莱州湾海咸水入侵可分为两种类型，一种为海水入侵，另一种为古咸水入侵。海咸水入侵的原因是当地地下水开发利用程度在枯水年份已超过当地地下水承载能力，也就是主要是农业灌溉用水量造成的采补不平衡，从更深层次的原因看是当地水资源已不能支撑社会经济的发展，即是一个社会经济问题。它不但涉及海咸水入侵区农业灌溉用水问题，关键是几十万农民的生存和发展问题，可以说，海咸水入侵问题，是农民用水管理不当的问题，农民用水管理的问题本质上讲是农民生存和发展问题。它是一个面上的问题，农业是几十万农民的事业，不是少数人能解决的事，控制和治理海咸水入侵，领导者和科技人员是非常重要的，但是他们必须同农民结合，才有作为。我们研究成果非常重要，但是要转化为农业生产力才能解决海咸水入侵区的农业问题，推动农业发展。目前，我国《水法》规定从江河、湖泊、河流和地下取水均要交水资源费，但至今农业取用地下水不收水资源费，也就是滨海平原区工业和生活取水易管理，而农业灌溉取水处于无序管理状态，枯水年、枯水灌溉季节各家各户为了满足作物灌溉需水超采地下水，缺乏节水和保护生态环境的意识。不解决农民用水管理、生存及发展的问题，仅从工程技术上去解决，海咸水入侵很难控制。因此，如何两手一齐抓，既如何充分利用多余洪水，拦蓄补源，又如何管理海咸水入侵区农民地下取水，如何使当地经济得到发展是关键课题。

3　农民用水者协会是农民自我管理用水防治海咸水入侵的有效途径

为了加强农村水利工程管理，实行测水、量水、水费征收和控制地下水开采制度，推广农民用水者协会，对促进节约用水，缓解海咸水入侵区水资源紧缺的矛盾，保证海咸水入侵区社会经济和环境的可持续发展，具有重要的意义[3]。在海咸水入侵区建立农民供水者协会目的，不同于其他地区，它不仅是解决工程运行费、维修费和折旧费问题，达到成本水价；更重要的是将农民组织起来，共同参与，通过有效措施，保护地下水不超采，就是保护自己子子孙孙生存的耕地不受破坏。

4　海咸水入侵区农民用水者协会的内涵、特点

农民用水者协会是由灌溉供水区、饮水供水区的受益农民自愿参加的群众性用水管理组织。经当地民政部门登记注册后，具有独立法人资格，实行独立核算，自负盈亏，实现经济自立。农民用水者协会负责所辖灌溉系统及供水系统的管理和运行，保证灌溉工程、供水工程资产的保质和增值。农民用水者协会由用水小组和用水协会会员组成[4]。用水小组是将供水受益区内的用水户按地形条件，水文边界，街道分布情况，同时兼顾村、村民小组边界划分的若干个用水单位。通过用水小组大会选举的每个用水组的用水者代表，即是该组的管水员，负责执行用水者代表大会作出的决定，管理本组的灌溉设施、供水设施

和用水工作，向本组用水户收取灌溉水费和饮用水费，并负责本组的其他日常工作。用水协会会员是指在协会范围内的受益农户户主，由用水户自愿申请加入，并经用水者协会批准。其权利和义务主要是：参加本组大会，有灌溉用水、人畜用水的权利和按时缴纳水费的义务，并根据要求参加保护、维护协会管理范围内的灌排设施、供水设施和地下水采补平衡，即不超采。

农民用水者协会的特点：它是具有法人地位的社会团体组织，在民政部门注册登记，取得合法地位，依照国家的法律、法规及协会章程运作，并对管理设施、财产和灌溉服务、人畜吃水供水服务负法律责任。它不是以营利为目的的服务性经济实体，通常按行政村或村民小组组建，不受当地行政机构的干预。农民用水者协会是农村从事灌溉服务或人畜用水服务的民主组织，用水户按自己的意愿民主选举协会的组织机构，内部事务通过会员代表大会以民主协商的方式确定，活动受到用水户的广泛监督。协会受政府业务主管部门的服务监督和指导。

存在的问题：打井和管道是农民自己建的，又不是项目投资的，如何才能组织农民管理自己；仅仅提供给每个农民用水者协会一些技术设备如水位计、电导仪，乡镇水利站是否可发挥出作用，例如定期发布水位和水质监测信息，包括警戒水位、破坏水位等。另外，提出了与农民用水者协会相关的研究内容和建立农民用水者协会的初始费用。

5 海咸水入侵区农民用水者协会建设的指导思想和内容

坚持以人为本，树立全面、协调、可持续的发展观，以保护海咸水入侵区水生态环境为目标，结合多余洪水利用工程和种植业结构调整，统筹规划，突出重点，建立海咸水入侵区农民用水者协会和示范区，改善农民的生存环境、生活和生产条件，促进农村脱贫致富。

建立农民用水者协会[5]，实质上是思想观念的转变，是建立在广泛宣传的基础上，通过对协会人员、用水者代表和广大用水户做大量的宣传，使之家喻户晓，深入人心，按自愿、公开、民主、因地制宜、依法、经济自立、水文边界与行政边界相结合的原则建立的。海咸水入侵井灌区一般以一眼机井为基本单元组建用水小组，由农民推选用水者代表；以一个电力变压器控制面积为单元，成立用水者协会。也可以自然村或行政村成立用水者协会，通过民主选举推荐用水者协会主席和执委会成员。即用水者协会的组织模式为"用水者协会+用水小组+用水户"。用水者协会主要是对计划用水进行监督，协助政府加强地下水资源管理，指导用水小组开展节水灌溉和技术服务。各用水组为独立核算的经济组织。规划在海咸水入侵区成立的单个用水者协会，大田灌溉面积一般控制为 2 000~5 000 亩，不宜过大，以便管理。

进行与农民用水者协会相关的技术支持的研究。由于海咸水入侵区农民用水者协会的建立不仅仅是用水管理，更重要的是如何使地下水在枯水年或连枯年或枯水季节不超采，达到保护生态环境的目的。通过研究划定海咸水入侵区地下水开采的分区，如禁采区、限采区等。选择适当的仪器和设备，监测地下水水位、水质，确定地下水开采的警戒水位、破坏水位等。

6 资金补助计划

建立农民用水者协会的费用支出主要为开办费、培训考察费、安装测水、量水设施，购买测量水位、水质的仪器等。

7 讨论

（1）在海咸水入侵区井灌区建立农民用水者协会面临着机遇和挑战。目前中国的农业生产是以家庭为最小生产单位，人均耕地一亩左右，土地经营规模较小，实行家庭联产承包责任制，村级行政管理为基础小农生产方式。农民追求每家、每户经济效益最大，他们重点关心自己收成好坏，能在作物干旱的时候灌溉，抽取地下水的电费省一些，而对能否节水，保护大家的共同家园不受破坏关心得少，但是一旦生态环境被破坏，他们希望政府或集体给解决问题。国家规定农民土地承包 30 年不变，村级政府具有分配、调整土地的权利。随着各地取消农业税，农民享受更多的利益，有利于农民用水者协会的建立。海咸水入侵问题涉及千家万户农民，是一个面上的事，也是随降雨量多少，不断变化，仅靠工程措施，很难解决好这些问题。将分散的农民通过农民用水者协会组织起来，在外界一定的技术指导和援助下，自我管理，而不是完全依靠村级政府，当然完全离开了村级政府，要想搞好水管理也不行。

（2）通过科学调查研究，依靠农民用水者协会，与农民协商，减少开采机井数量，达到合理布局。自 1976 年后，莱州市滨海平原机井密度大幅度增加，农业产量大幅度提高，地下水开采量也同样大幅度增加，结果造成海咸水入侵这种灾难性后果，因此，利用协会的力量，公平合理的封闭部分机井是有效的途径。

（3）在海咸水入侵区，通过农民用水者协会，示范推广农业灌溉节水技术包括农艺节水措施，调整农业种植结构、选用耐旱、耐盐作物品种等。

（4）通过农民用水者协会，将农民组织起来，配合地下水回灌补源工程的实施。由于农民用水者协会是农民自愿参加的社会团体组织，能否承担保护自己家园又能发展生产的责任，只有先试点，不断完善，取得经验后，再推广。

参考文献

[1] 王秋贤，任志远，孙根牛. 莱州湾东南沿岸地区海水入侵灾害研究[J]. 海洋环境科学，2002（3）.
[2] 王维平，曲士松，孙小滨. 基于水资源承载力的农业结构和布局调整分析[J]. 中国农村水利水电，2008（12）.
[3] 曲士松，王维平，孙小滨. 海水入侵区农民参与水生态修复管理初探[J]. 灌溉排水学报，2007（5）.
[4] 胡登贵，吴化南，程荣. 建立灌区农民用水者协会的探讨[J]. 中国农村水利水电，1996（3）.
[5] 杜杰，孙海燕. 发展村级农民用水者协会实践中的几个问题[J]. 水利发展研究，2007（9）.

山东省滨海地区缺水造成的社会经济问题

林洪孝
山东农业大学,泰安

1 山东省滨海地区缺水原因分析

山东省滨海地区缺水原因主要包括以下几方面：①水资源匮乏；②水质污染；③经济快速增长需水量大；④人口聚集（城市化）；⑤管理不善和缺乏节水意识；⑥蓄水工程不足。

据《山东省城市供水水源规划》，位于山东滨海地区的烟台、青岛、威海、日照等城市都表现出很大程度的缺水，据预测 2020 年缺水率（保证率 75%）烟台将达到 32.34%、莱州市 81.06%、招远市 58.20%、龙口市 15.79%。

若以山东半岛测算，随着半岛地区社会和国民经济的迅速发展，各部门的需水量与日俱增，使得水资源的供需矛盾日趋尖锐。按照现状用水水平：城市生活用水 83~193 L/（人·d），工业用水 66~139 m^3/万元，农村生活用水 45 L/（人·d），农业用水 66~139 m^3/亩，菜田用水 280~412 m^3/亩，牲畜用水 20~40 L/（头·d），林业用水 100 m^3/亩，坑塘养殖 800 m^3/亩，分析计算得现状年半岛地区 50%、75%保证率的社会总需水量分别为 78.03 亿 m^3/a 和 91.82 亿 m^3/a。与现状工程情况下的可供水量进行平衡分析，得相应上述两种保证率年份的缺水量分别为 24.17 亿 m^3、44.29 亿 m^3。

2 滨海地区缺水造成的社会经济影响

自 20 世纪 80 年代以来，山东省连续干旱，地表径流大量减少，为满足社会经济发展需水要求，只有加大地下水的开采，以弥补地表水的不足，但受区域水资源先天不足的影响，仍难以满足社会经济建设和发展的需水要求，缺水形势仍日趋严重，并产生了一系列严重后果。

2.1 超量开采水资源，形成地下水降落漏斗，削弱河流稀释能力

据历年监测资料分析，烟台市夹河中下游的芝罘区三水厂、烟台电厂水厂、芝罘区五水厂的开采漏斗，已由原来的单一型发展为多个漏斗相连，形成较大范围的开采漏斗；潍坊市地下水漏斗面积逐年扩大，在北部河流冲洪积平原区形成了串珠状分布的漏斗带；威

海市集中开采地下水的环翠区田村镇涝台水厂已形成降落漏斗；滨州市在博兴地段形成东西向长条形漏斗，并与广饶漏斗、桓台漏斗连成一片。

有的含水层被开采疏干后，还造成了地面塌陷，如大沽河堤防因此造成多处裂缝和塌坑，严重危及工程安全，有的房屋裂缝。

同时，由于大量开采地表水、地下水，使天然径流减少，降低了水环境的纳污能力，削弱了产生和容纳水的潜力。

2.2 引发海水入侵，造成严重区域水文地质环境灾害

滨海平原地区已出现了大面积地下水低于海平面的负值区，致使海水入侵。20世纪80年代末期，龙口、莱州等沿海地带，海水和咸水入侵面积594 km^2，1995年莱州湾地区海水入侵面积已达到970 km^{2}以上。海水入侵淡水层，直接导致地下水环境逐渐恶化，降低了地下水的可用性，并在一些河流沿岸地带造成污水大量下渗，地下水水质受到严重污染，加剧了水资源的短缺，给人畜饮水和工农业生产带来了严重困难和巨大经济损失。

2.3 影响人民生活和生活质量的提高

现在的海水、咸水浸染区，过去地下水一般都比较丰富，人畜吃水不成问题，可是在地下水被浸染后，人畜吃水发生了困难。据不完全统计，浸染区目前已有45余万人吃水发生了困难。为了解决浸染区的人畜吃水问题，政府和当地人民花费了大量资金，采取了很多措施，如广饶、寿光、昌邑等市的有些浸染区，过去吃水井只有10 m多深，浅层水被浸染后现在只能打100 m或200 m以上的深井，提取深层淡水来解决。潍坊市的寒亭区及寿光市，浅层水被浸染，深层水含氟量又高，饮用后当地氟病增加。潍坊北部地区一部分村镇不得不靠峡山水库定期放水来解决吃水问题，近几年水库无水，就只能喝咸水。还有的地方当地水解决不了，只得在外地打井，远距离引水来解决人畜用水问题。如崂山区白沙河入海口左岸有5个村庄，1964年在村附近打井供水，1970年供水井变咸报废，又在上游500 m以外打井供水，后又变咸报废，现在不得不在上游2 km以外的流亭镇附近打井供水。再如青岛市城阳附近有6个村庄因海水浸染无水吃，不得不集资70多万元从上游的虹字河水库引水，由于虹字河水库仅蓄水109万 m^3，供水没有保证，每年还需从书院水库引水10万 m^{3}补充虹字河水库，然后再由虹字河水库引水入村供人畜饮用。

浸染区的供水问题，虽然各地采取了不少措施，但仍有一些村庄没有解决，目前利用肩挑、人抬、车拉、驴驮到外地运水吃的地方还为数不少，供水问题亟待解决。

海咸水浸染区，由于生活用水发生了困难，有些地方不得不饮用劣质水，使某些疾病的发病率增高。据对莱州湾沿岸的龙口、招远、莱州、平度、昌邑、寒亭、寿光七个县（市、区）的统计，共有患甲状腺肿、氟斑牙、氟骨病、布氏菌病、肝吸虫病症的人数达45.1万人。有些病种过去就存在，有些病种是由于地下水被浸染后引进或加重的。如潍坊北部三县（市、区）浅层水变咸，只能饮用深层含氟淡水，结果使患氟病人数猛增，仅寿光市患各种氟病人数就由20万人增加到40万人。莱州市1978年地方病普查时，氟斑牙病患者为3076人，至1986年发展到1.5万人，12年增加了近5倍。

2.4 降低水保障社会经济建设与发展的能力

进入 20 世纪 80 年代以后，由于遭受连续干旱的影响，使原本就十分紧缺的当地水资源变得更为稀少，除潍坊、青岛在 1989 年末至 1990 年初，通过"引黄济青"工程补充黄河水 2.7 亿 m^3 外，其他地区均无客水接济，因此，属缺水最严重的地区。1980—1989 年农业生产平均成灾面积为 810 万亩，占耕地面积的 28%，占山东省受旱成灾面积的 35%，1989 年伏旱严重，仅烟台、威海两市的绝产面积就达 250 万亩，占当年耕地的 25%。尽管 1990 年降雨比较丰沛，农业生产有所回升，但是潍坊、青岛、烟台、威海四市 1980 年至 1990 年的粮食总产递增率平均只有 2.65%，低于山东省平均递增率 4.12%；花生总产同期年递增率 2.80%，亦低于山东省平均 4.14%的增长水平。烟台市 1980 年粮食总产 265.2 万 t，而 1989 年总产却降至 239.34 万 t。

工业生产也因供水不足而造成严重损失，据 1982—1984 年的不完全统计，青岛市因缺水减少工业产值 14.05 亿元，影响利税 3.08 亿元。供水最困难的 1989 年，仅烟台、潍坊两市就有 151 个企业因严重缺水而被迫限产或停产，减少工业产值 21.5 亿元，影响利税 2 亿元以上。莱州市 79 个市属以上企业，其中有 18 个处于海水浸染区，据不完全统计，每年的工业损失达 1.5 亿元左右。

多年的实践已证明，供水不足已成为当前而且也将是今后制约该地区国民经济发展的主要因素。

2.5 造成生产与生活、地区间用水矛盾日益尖锐

由于该区域社会经济的较大规模的快速发展，需水量增长较快，原有许多农业用水水源转供城市和工业用水，尤其是水库的水，加剧了城乡之间、生产与生活之间，乃至地区之间和生态环境之间的水资源供求关系，并演变出恶性循环的争水局面。

3 建议

3.1 应进一步调研该区域缺水的性质和深度，并建立相应的评估方法和指标体系

滨海地区各地水源条件、供水工程条件、社会经济建设与发展的结构、规模、速度等情况均有一些差别，应进一步进行详细的深入调研和细化分析，并建立一套评价缺水及其对社会经济建设发展影响的评价方法及指标体系，以比较全面、合理地反映该区缺水和影响的情况。

3.2 运用循环经济和生命周期理论建立区域水运行全过程评价方法

水资源的开发利用和管理保护应针对水资源环境、水源、供水、用水、节水、排水及处理回用的全过程进行评估分析，方可提出有效率和效益的综合性方案，运用循环经济理论和生命周期理论是比较适用的，它不仅能够系统反映水的自然循环规律、社会经济用水状况和特点、特征，还能够促进对水的有效利用和管理。

3.3 深入研究水对区域社会经济发展的作用和影响

水是基础性的自然资源，经济性的战略资源。对于该地区应定量研究水对环境良性循环、社会经济的建设与发展的影响和限制性作用，尤其是用经济学的观点分别定量研究对生活、第一、第二、第三产业的保障和制约作用的大小，并建立相应的评估指标体系，深入反映水对社会经济发展的作用。

3.4 构建节水型社会管理模式和节水型社会生产与生活评估标准指标体系

建设节水型社会无疑是水资源开发利用和管理保护的发展方向，但对于特定地域的滨海地区来说，建设节水型社会必将面临如何协调处理比较脆弱的水生态环境安全与提高人民生活质量、社会经济快速发展需水之间的矛盾，因而应构建适应社会发展的水管理模式、制度和运作机制。在资源管理与产业开发利用管理相对分离的前提下，引入水务一体化的管理模式和资源水权、排水权，以及水设施特许经营权等的市场化运作方式应是发展方向。同时，对水的利用和管理应建立相应的考核指标体系和有效的管理制度。

3.5 研究提高水保障社会经济建设与发展的投入产出模式和管理运行机制

水利用效率、效益不高，管理粗放，用水技术水平还比较落后是当前面临的重要问题，因而深入研究区域水的投入产出模式和管理运作机制尤显重要和紧迫，它是研究水对区域社会经济作用的重要内容，也是提高水资源开发利用水平和管理能力的需要。

3.6 加大对该区域海水入侵规律和特点的研究力度，进一步提出防止海水入侵的理论和技术方法

海水入侵是滨海地区共同面临的威胁，在总结以前研究成果的基础上，进一步深入研究该地区海水入侵的规律和特点，是防止和治理海水入侵的前提，更是提出切实可行的水资源开发利用和管理方法的需要，对此应加大研究力度。

3.7 研究地表水的多途径开发利用方式和进一步提高水利用效率的技术方法

该地区雨水、洪水资源相对比较丰富，污废水的利用率还比较低，研究多途径的开发利用方式是非常重要的，这既是解决当地缺水的需要，更是抑制和治理超量开采地下水的需要，它比大规模从外区域调水要经济得多，并更有利于当地生态环境的建设。

4 结语

缺水对山东滨海地区的生态环境和社会经济建设与发展造成了很大的影响，已成为当地进一步发展的制约性要素，深入研究该问题是非常重要和必要的，并将对相关课题的研究和开展起到促进作用，有利于提高水利用效率、效益和节水型社会的建设。但是，由于对一个确定地区缺水问题及其影响的研究，不仅涉及的内容多而复杂，更需要深入当地做细致的调查研究，方可提出解决问题的办法。这也是本文提出建议的原因，请大家参考。

王屋水库饮用水水源污染风险性评价

李新军[1,2]　张保祥[2]　孟凡海[3]
1. 山东省济南大学资源与环境学院，济南
2. 山东省水利科学研究院，济南
3. 山东省龙口市水务局，龙口

水是生命之源，饮用水水源更是与人类的生存息息相关。随着我国经济的快速发展，污染物排放量逐年增加，水环境污染问题日益严重，特别是饮用水安全受到威胁，水源地的保护和饮用水的健康愈加重要[1]。

基于风险的水污染控制管理已成为目前世界上许多国家广泛采用的模式，它突破传统的环境质量标准管理模式，以保护人体健康和生态安全为最终目的[2]。目前国内在水资源污染风险性评价方面的研究多为地下水污染风险性评价和脆弱性评价研究，关于地表水污染风险性评价的研究很少[3-4]。水资源污染风险性的概念是水资源脆弱性概念的延伸，即是在水资源脆弱性研究不断深化的基础上发展而来的。相对于地下水污染风险性的概念[5]，地表水污染风险性是指地表水由于人类活动而遭受污染到不可接受的水平的可能性。这是地表水资源固有脆弱性与人类活动造成的污染负荷之间相互作用的结果。因此，在地表水污染风险评价中不仅要考虑地表水资源纳污的能力，也要考虑研究区不同人类活动所产生的污染负荷。

地表饮用水水源作为一种关系人类身体健康的水资源，对其进行污染风险性评价更加具有必要性。由研究区以往污染物研究资料、现场调查等得到的地表饮用水水源脆弱性表征了人类活动造成的污染负荷；由水源区各项特征综合评价得到的地表饮用水水源敏感性评价则反映了地表饮用水水源纳污的能力。将地表饮用水水源敏感性评价和脆弱性评价结合起来，便构成了完整的地表饮用水水源污染风险评价框架体系。评价思路如图1所示。

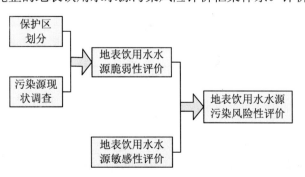

图1　地表饮用水水源污染风险性评价思路

1 脆弱性评价

1.1 保护区划分

地表饮用水的安全关系着人类的生存与健康,在防治水污染方面有着更高、更严格的要求。在进行地表饮用水水源污染风险性评价时,首先要对水源地进行保护区划分。饮用水水源保护区划分的主要依据是国家环境保护总局 2007 年发布的《饮用水水源保护区划分技术规范》(HJ/T 338—2007)。饮用水水源保护区一般分为一级保护区、二级保护区和准保护区(亦称三级保护区)。目前国内地表饮用水水源保护区的划分主要有类比经验法和数值模拟法[6-8]。

1.2 污染源调查

地表饮用水水源研究区内现有的和潜在的污染源主要是通过两种方式调查得来并存档:一是研究区有关污染源的历史资料和条例法规;二是现场调查。调查的污染源包括农业污染源、商业污染源、工业污染源、市政生活污染源等。调查完污染源以后,将这些污染源名称、污染源位置、污染源描述、可能造成的污染等信息按污染源类别分类存档,为后期工作做好资料准备。

1.3 脆弱性评价

地表饮用水水源脆弱性评价的决策因子有四个:一是污染物化学特征。二是污染物排放、泄漏及污染事故发生的可能性。三是污染物监测历史及周边环境纳污能力。四是潜在污染源的数量及密集程度。

潜在污染源的数量及密集程度是脆弱性评价很重要的一个方面,而且其资料容易收集且客观性强,因此,在进行地表饮用水水源脆弱性评价中,通常以潜在污染源的数量及密集程度评价为主。在对地表饮用水水源进行保护区划分与污染源调查的基础上,地表饮用水水源脆弱性分级如表 1 所示。

表 1 地表饮用水水源脆弱性分级

	一级保护区	二级保护区	准保护区	分级
分区内潜在污染源数量	≥1	≥5	≥7	高
	0	3~4	4~6	中
	0	1~2	1~3	低

2 敏感性评价

地表饮用水水源敏感性评价的决策因子有三个:一是河流流速或水库大小;二是引水方式及设备完善性;三是 WRASTIC 指标体系。

2.1 河流流速或水库大小

较小的流域或水库在遇到突发的污染事故或污染物排放时将受到更大影响。因此，在进行水源敏感性分析时要考虑相关水库大小以及河流流速。根据以往研究经验，对水库大小及河流流速进行敏感性分级，分级结果如表2所示。

表 2 河流流速或水库大小敏感性等级

水库库容/亿 m^3	多年平均河流流速/(m^3/s)	等级
<0.1	<2	高
0.1~1	2~11	中
>1	>11	低

2.2 引水方式及设备完善性

引水设施、取水设备的完善性及各种设施的维护等对地表饮用水水源敏感性都有很大的影响。引水方式及取水口处原水日平均浑浊度能够比较科学合理地体现引水设施及其完善性的水平。在地表饮用水水源敏感性评价中，采取直接引水的其敏感性为高，采取间接引水的其敏感性为低；取水口处原水浑浊度大于 10 度的，其敏感性为高；浑浊度小于等于 10 度的，其敏感性为低。表 3 为比较取水方式及浑浊度得到的引水方式及设备完善性的敏感性分级。

表 3 引水方式及设备完善性的敏感性等级

日平均浑浊度等级 \ 引水方式等级	高	低
高	高	中
低	中	低

2.3 WRASTIC 指标体系

WRASTIC 指标体系是基于流域特征及土地利用等方面进行地表饮用水水源敏感性评价的方法[9]。WRASTIC 指标体系包括污水排放（W）、休闲用地影响（R）、农业用地影响（A）、流域大小（S）、交通道路影响（T）、工业用地影响（I）以及地表植被覆盖率（C）。WRASTIC 指标计算公式为：

$$V_{指标} = \sum R_i W_i \tag{1}$$

式中，$V_{指标}$——指标值；R_i——评分值；W_i——权重，i 取 W、R、A、S、T、I 和 C。

根据各项指标可能对饮用水水源造成的污染以及前人研究经验，WRASTIC 各指标的权重依次为 3、2、2、1、1、4、1。各项指标的评分值主要是根据现状调查资料来确定。经计算得到的 WRASTIC 指标值越大，地表饮用水水源敏感性越高。当 WRASTIC 指标值大于 50 时，其敏感性等级为高；当指标值介于 26～50 时，其敏感性等级为中；当指标值小于 26 时，其敏感性等级为低。

2.4 地表饮用水水源敏感性评价

在对以上三个决策因子进行易污性分级以后，再对每个决策因子赋值。决策因子敏感性等级分别为高、中、低时，敏感性值相应的分别为 10、5、1。地表饮用水水源敏感性总分为三个决策因子敏感性值之和。根据敏感性总值进行地表饮用水敏感性分级，如表 4 所示。

表 4 地表饮用水水源敏感性等级

总分	20～30	11～19	0～10
等级	高	中	低

3 应用实例

3.1 研究区概况

王屋水库位于山东省龙口市黄水河中上游，龙口市石良镇和七甲镇境内，控制流域面积 320 km²，总库容 1.21 亿 m³，兴利库容 7250 万 m³，是集城市供水、防洪、灌溉、养鱼等多功能于一体的大（二）型水库，也是龙口市最大的城市生活用水、工业用水水源地。近几年，随着社会经济的发展，龙口、栖霞、招远境内的企业将很多污染物排入黄水河上游，对王屋水库水质产生了一定的影响。作为龙口市唯一的城市饮用水水源地，保护王屋水库水质安全责任重大。研究区地理位置见图 2。

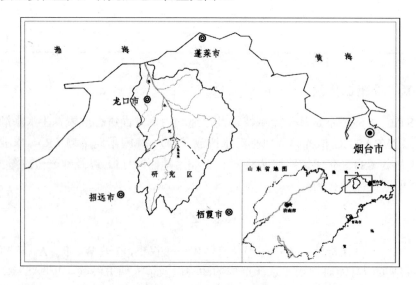

图 2 研究区地理位置图

3.2 评价结果

在对研究区相关资料研究及污染源现状调查的基础上，利用上述评价方法对王屋水库饮用水水源进行污染风险评价。

经过对研究区内潜在污染源的调查，发现王屋水库准保护区内存在十几个潜在污染源，这些污染源包括电解厂、制革厂、石材加工企业、金矿企业等。根据地表饮用水水源脆弱性评价方法，王屋水库饮用水水源脆弱性等级为高。

王屋水库库容为 1.21 亿 m^3，则其水库大小敏感性等级为低；王屋水库饮用水取水方式为间接引水，且取水口原水浑浊度为 4，则其引水设备及其完善性敏感性等级为低；通过对研究区现状调查，经计算得到的王屋水库饮用水水源 WRASTIC 指标敏感性等级为中。由此可知，王屋水库饮用水水源敏感性总分为 7，敏感性等级为低。

在对地表饮用水水源敏感性评价与脆弱性评价的基础上，综合分析二者评价结果，就可得到地表饮用水水源污染风险性评价结果。地表饮用水水源污染风险性分级如表 5 所示。

表 5 地表饮用水水源污染风险性分级

脆弱性 \ 污染风险性 \ 敏感性	高	中	低
高	高	高	中
中	高	中	中
低	中	中	低

根据上述评价方法，王屋水库饮用水水源污染风险性等级为中。

4 结语

通过对王屋水库饮用水水源污染风险性评价可知，造成王屋水库饮用水水源污染风险性为中的主要原因是王屋水库准保护区内潜在污染源数量较多。因此，在建设项目时，在饮用水源保护区内应尽量避免建设污染性强的工业企业，居民区尽量全部铺设下水道。

地表水饮用水水源污染风险性评价是基于地表饮用水水源敏感性评价和脆弱性评价进行的。但是由于对地表水敏感性和地表水脆弱性还没有统一的认识，特别是关于地表水敏感性的研究刚刚起步，还需要在实践中不断总结经验，完善地表饮用水水源污染风险性评价的理论和方法。

参考文献

[1] 倪彬, 王洪波, 李旭东, 等. 湖泊饮用水源地水环境健康风险评价[J]. 环境科学研究, 2010, 23（1）: 74-79.

[2] 王丽萍, 周晓蔚, 李继清. 饮用水源污染风险评价的模糊——随机模型研究[J]. 清华大学学报, 2008,

48（9）：1449-1452，1457.

[3] 刘绿柳. 水资源脆弱性及其定量评价[J]. 水土保持通报，2002，22（2）：41-44.

[4] 李凤霞，郭建平. 水资源脆弱性的研究进展[J]. 气象科技，2006，34（6）：731-734.

[5] 张丽君. 地下水脆弱性和风险性评价研究进展综述[J]. 水文地质工程地质，2006（6）：113-119.

[6] 贺涛，彭晓春，白中炎，等. 水库型饮用水水源保护区划分方法比较[J]. 资源开发与市场，2009，25（2）：122-123，185.

[7] 李建新. 德国饮用水水源保护区的建立与保护[J]. 地理科学进展，1998，17（4）：88-97.

[8] 史莉，尹璐，李越越，等. 贵阳市阿哈水库饮用水水源地现状及保护对策[J]. 贵州水力发电，2008，22（3）：20-23.

[9] New Mexico Environment Department. State of New Mexico source water assessment and protection program[N]. Santa Fe，2000.

城市水资源管理
——全球性挑战

W. F. 盖格
杜伊斯堡—埃森大学，德国

1 对水资源可利用性的错误认识

水是人们日常生活及工农业生产最基本的要素之一。很多地区的可更新水资源量十分有限，而深层地下水资源，通常认为其在短期内是不可更新的。即使如此，与其他自然资源相比较，人们对水资源的滥用却是有过之而无不及，似乎水资源取之不尽、用之不竭。哲学阐释及科技进步都在某种程度上促成了人们对水资源无限存在性这一错误认识的形成。

回顾人类文明的进步历程可以发现，城市大都选择依水而建。水可以为人们直接利用，有助力于商业贸易进而为人们带来财富和各种福祉。早在公元前的米莱托城（Miletus），由于无法对地下水进行开采利用，人们就已经把可更新水资源的可持续管理与艺术及建筑很好地结合起来，并成功地运用到供水及污水处理过程当中，很好地满足了社会的需求。米莱托的哲学家泰勒斯（Thales，公元前 624—公元前 545 年）最早提出了水循环的思想，柏拉图（Platon，公元前 427—公元前 347 年）后来又对其进行了讨论，直到两千年后才由帕利西（Palissy，1510—1590 年）对水循环做出了科学的解释。水的无限循环理论告诉我们雨后总会有水可用，而河流之中总是流淌着从未知的上游某地汇集而来的水流，喷涌的泉水总会源源不断地把地下水带到地表供人们使用。

工业革命后电动水泵的诞生为人类的用水带来了极大的方便，即使在干旱季节人们也可以利用电动水泵抽水使用，从而水资源可持续管理的观念便被人们逐渐淡忘了。现在，凡是公共供水系统覆盖到的地方，人们只要拧开水龙头就会有水源源不断地流出来，人们对水资源随自然条件的变化而时刻变动的认识也消失的无影无踪了。即便在公共供水系统覆盖不到的偏远农村地区，人们也可以开采利用地下水，近年来开采深度也在逐渐加深。由于供水在很大程度上得到保障，很少有人再去费力在雨季进行雨水收集。相反，当洪水发生时，雨水总是被想方设法尽快地排走。

综观上面所提及的水循环的思想、无限重复的降水过程、持续涌流的泉水、奔流不息的江河以及日常生活中触手可及的自来水供应，都给人们留下了水资源取之不尽、用之不竭的印象，这种认识从人们日常的用水习惯中可以清楚地反映出来，尤其是在水资源匮乏地区，渗漏或根本不关的水龙头并非罕见。水成为一种供人任意使用的物品，没人关心水从哪里来，水又将会到哪里去。

从某种意义上来说，当前正在上演的工业化及城市化进程是在牺牲后代人利益的基础上进行的。发达国家走过的"先污染后治理"的发展道路是人类社会进程中所犯下的最大错误之一，因而发展中国家没有理由再去重复这种牺牲环境换取经济进步的发展模式。更糟糕的是，发展中国家的城市扩张一般先于城市基础设施的建设，仅有少数城市可以提供足够的人力及物力资源来满足快速增长的用水及卫生需求。贫民区内恶劣的环境状况更是社会性抑郁及各种犯罪滋生的温床。此外，水的问题还涉及政治问题，在城市向外扩展的过程中，水的问题总是在道路、能源及通信系统完善之后再加以解决的。而更严格来说，水才是城市扩展及规划赛程中应该最优先考虑的要素。有预计到2025年，世界人口的60%都将居住在城市地区，南美、非洲及东南亚的城市群规模也将加倍，届时城市水资源形势将更为严峻。

2 前所未有的挑战

2.1 水资源匮乏

作为人类文明的中心，城市同时也是各种用水活动的中心。如果当地水资源比较有限，城市供水水源会延伸到周边甚至更远的地方。通常城市规模越大，这种对整个区域内水资源的主宰作用也就越明显。在20世纪，北京市内的清洁水被用于冷却、洗车及街道除尘，草坪灌溉也使用了大规模的喷灌设施，而同时市内的雨水却通过粗大的管道排出城区，在此期间，北京地区的地下水位以平均每年2 m的速度下降，这种强烈的对比实在令人费解。同样令人费解的是在水资源匮乏的地中海地区，夏季每名游客的用水量可以达到1 000 L/d。类似的浪费水的行为在世界各地随处可见，当前全球城市化进程中城市的快速膨胀更是加剧了人们对水资源的浪费性使用。

在城市向外扩展的过程中，原先的绿地被建筑、道路、庭院及广场等硬化地面逐渐取代，加速了城区地表径流的形成，同时也限制了地下水的补给。为确保市民的安全，雨洪水被快速排入下水管网及城市河道，最终流出城市，因而不可避免地造成当地水资源逐渐减少，同时也对水文循环形成不可逆转的干扰。在城市向外扩展的初期，这种过度用水及城市下垫面变化的后果并不轻易为人所察觉，因为随着技术的进步，各种新的资源得以开发利用，如深层地下水资源。然而由于对地下水的过度开采，城区地下水位迅速下降的报道已经常见诸报端。

2.2 水污染

人们不假思索地将各种污染物排放到土壤及水体当中，导致土壤及水体不能再为人们所用。通常，借助于地表水及地下水的运动，这些污染问题被转移到其他地区或最终转移给海洋。而那些埋藏在土壤及沉积物当中的化学"定时炸弹"的危害还很少为人所知，并将进一步转移给后代。土地利用方式及气候的变化很有可能将这些潜在的有毒物质再度激活，并威胁到人的生命安全。

2.3 水价

鉴于政治因素，水价的征收经常低于供水成本，因而用户很难意识到水的真实价值，这在某种程度上更鼓励了人们用水浪费的行为。以印度某社区为例，该区水资源极其缺乏，政府投入很大成本从外地调水，但并不向用水户计收水费。当有人问为什么在没人用水的时候公共水龙头也从来不关时，得到的回答竟然是"水又不值一分钱"。

2.4 当前水管理理念的局限

现实生活中由于过度用水造成的相关问题，很难及时被人察觉，即使意识到了问题的存在，其造成的压力也未必能让当局意识到对其进行认真处理的必要性；再进一步，即使当局对其认真对待了，也未必能够进入最终决策的范围；最后并非所有的决策都能够最终付诸行动。当前常用的末端治理的理念通常都是针对各种问题所造成的直接危害，远远触及不到问题产生的源头；当前的某些治水理念中，水甚至被当成敌人来对待，例如，雨洪水及使用过的水要尽快排出城市并在城市外部对其进行处理。治水的各种措施也是头痛医头、脚痛医脚，不成体系。在这种管理理念下，不可避免地会出现地下水位下降、洪涝灾害频发及生物消亡的现象。

2.5 新出现的问题

世界上很多地方出现了一种新的现象，一些超大城市呈集聚发展态势，如香港、深圳和广州之间的珠江三角洲，雅加达—泗水走廊（印度尼西亚）及日本位于东京、长屋王及大阪之间的东海道走廊。前面所提到的各种问题在这些城市群内尤其突出，很明显传统的供水及污水处理方法已不能够解决这些问题。城市集聚发展带来的新问题，再加上20世纪遗留下的老问题，包括世界上有10亿人的饮用水得不到保障，20亿人没有足够的卫生设施，40亿人用水产生的污水得不到很好的处理，都给当前的形势造成了极大的挑战。不管承认或不承认这些问题是否是人为因素，或忽略水资源不充足带来的后果，这些问题都将依然存在。

气候变化也将对淡水资源产生很大的影响，由此造成对人类社会及生态系统更为多样的后果。据估计，未来降水的强度将会增加，分配不均也将进一步加剧，许多地区发生洪涝及干旱灾害的风险加大。由于这种洪水及干旱等极端现象的出现，各种形式的水污染的危害将会更加严重。当前的水管理实践很有可能难以应对以上所提到的由气候变化所带来的诸多挑战（Bates et al., 2008）。

2.6 将社会及生态因素纳入水管理实践的尝试以失败而告终

除了用于饮用及生产用水外，城市水体还有其文化、审美、社会及生态功能。早在1970年，早期的城市水文学者McPherson教授就已申明在水管理过程中要充分考虑到社会和经济方面的影响因素。当前进行的各种相互孤立的技术努力，根本不可能达到既技术可行又经济有效的可持续发展目标，很显然社会稳定及生态平衡也将受到威胁。此外，若就业及社会结构变动太快，则可能会使已经取得的成果遭到破坏，最终导致新的失业率增加，更无力处理前工业化所遗留下来的问题。

3 呼吁分散式水管理方案

3.1 亟待水资源综合管理

决策者们已渐渐地意识到了经济的发展与环境状况及当地水资源量多少之间密不可分的关系，忽视水资源的承载能力将会限制经济的增长，甚至破坏已经取得的成果。从这个角度来看，贫困既是造成城市水问题的主要原因，也是其导致的主要后果。

随着全球化的不断深入，各大城市也在努力寻求新的工业和经济发展机会。然而经济的发展与水资源可利用量及供水成本紧密相连。可持续发展需要改变人们的发展观念，发展中所耗水量需视当地水资源可利用量而定，供水不应以牺牲环境为代价。因而决策者们必须搞清楚基本需水与附加需水之间的区别，古罗马时期的尼姆城就是这方面的绝佳典范。尼姆城的供水系统分别供应公共用水、商业用水、洗浴用水及私人住宅用水，分水池的独特设计确保了公共用水优先得以供应，只有当公共用水得以满足后，剩余的水才会用于供给其他用水。

3.2 让基层水管理部门活跃起来

好的水管理措施要适应本地自然条件，尤其要考虑水资源可利用总量这个前提，兼顾社会及经济发展需求。在进行与水相关系统开发设计时没有固定的模式可循，解决每座城市、城市里的每片区域甚至每个社区的水问题都要对症下药，一座城市的治水经验或许并不能用于解决其他城市的水问题。因而，必须让基层水管理部门采取更多的行动，承担更多的义务。相对于大型集中式供水及水处理系统而言，小型自给自足式系统的成本较低，操作风险也较低，而且小规模建设及运行也有利于稳固社会经济结构。在小型系统的运行中，系统负责人及用户对该社区及区内设施有归属感，因而更愿意为系统的稳定运行承担更多的义务。但如果一个社区是不同文化、宗教及社会阶层的聚居区，则很难让人们找到这种归属感。

3.3 行政体制改革

当前采用的传统水管理机构设置模式不利于分散式水系统的引入。集中式水管理机构设置通常根据其具体职能划分为灌溉、供水、排水及污水处理等多个部门，高层往往只对给排水比较重视，因而忽略水资源的其他各种用途。供水中对各种不同用途用水并不加以区分，往往以最高水质标准供水，导致供水成本大增。在一定程度上，当前的水务行政管理主要着眼于集中式系统。然而，实践已经证明，无论从经济还是可操作性角度来看，相对于集中式系统而言，半分散式甚至是分散式供水与排水系统都有着很大的优势。因而水务管理行政体制也需要加以改进以适应新的分散式系统模式。

3.4 充分利用本地资源

对城市供水提倡采用分质供水系统，并充分发掘本地水资源潜力，如充分利用雨水及中水进行灌溉及景观水池补水，这也有助于改善当地的生态环境。供水要优先考虑当地水源，降低各社区对外界水源的依赖性。各社区需找准其自身的社会定位，建立起适合自己传统及文化需求的用水模式。通常，市民们在防范未来可能出现的各种社会问题方面比政府组织更具创造性，因而应积极鼓励群众参与到治水决策中来，以发掘并充分利用本地智慧。

3.5 超大城市的水系统必将走向分散式

即便在超大城市内，分散式水系统仍有其生存空间，因为人们的日常起居生活仍是以社区为单位，购物及文化活动也大多在社区内进行。因而无论城市规模大小，人们的活动在很大程度上始终局限在一定的空间范围内。未来的城市发展格局仍将是以社区为单位，并取决于小范围内的生活方式及发展模式。高层水务管理机构只需对基层商业性给排水运营商进行监管控制即可。当然这个理念无疑又会引来人们对"规模效益"和"小型系统更稳健灵活"各自利弊的一番讨论。

问题的关键在于这种预期的变化需要多长时间才能变成现实。公元前 800 年时赫西俄德（Hesiod）指出水污染与人类健康之间的基本关系，但直到 300 年后古希腊的城市中才得以安装了各种卫生设施，尽管类似设施在早前的美索不达米亚平原（Mesopotamia）及印度河流域就已经存在了。这种变化的发生其实也是一个学习的过程，首先从学校教育开始，一直贯穿人的一生。对水的认识的转变也必须从用户开始。

4 从本地化行动到全球式思维

从某种程度上来讲，大尺度内各地的水资源相关问题有许多相似之处，水文学基本原理及各种相关技术到处都适用，因而针对各地问题量身定做的本地化解决方法并不排除全球化思维的指导。即便对于局地分散式水系统，也可制定出一些具有普遍适用性的指导方针以及投资小效益大的水管理措施建议，节水措施也是处处通用。若分散式水管理得以实施，世界范围内需水量增长的趋势将有望得到遏制，管理不善的现象也有望得到改善。从全球角度来看，以下的指导方针普遍适用，首先所有的用水户都必须被吸纳到水管理实践中来，流域不仅是水文单元，同时也应该当做经济单元来看待，流域水管理方案制定时也应该给基层留下一定的自由空间，水费的征收必须要体现出水的真实成本。

自然界生态系统的划分不会以人为的国界为界，因而水资源管理中经常需要进行国与国之间的相互合作。在国际合作的框架下，同样也应该给区域及局地的水管理留下一定的自由空间。全球化并不代表标准化，从全球尺度来讲有必要对一些基本准则达成共识并加以遵守，但从局地角度来看，每个区域又必须找出最适合自己的水管理方法。

全球性思维倡导相互合作共同治理水资源，但在这方面发达国家理应做出表率，承担主要责任。由于发达国家较发展中国家而言，经济实力更为雄厚，应首先引入新的水管理机制。其实发达国家过去是在以自然环境为代价的基础上发展起来的，因而其有义务向人们展示一条更好的发展道路，仅仅表明此类意愿对水资源管理毫无用处，或仅仅向发展中国家提供资源保护的建议一样无益于水资源管理。更何况发达国家在 20 世纪下半叶所遭受的环境危机至今也尚未完全解除，若采用新的水管理方式，其自身也将从中受益。从这个意义上来看，全球城市间的经济战也必将转变成争夺优越环境的生态战，并最终形成更具竞争力的经济结构，促进可持续发展。

5　展望未来

　　我们希望当尊敬的读者站在 2030 年回望时，我们现在的展望届时将成为现实：20 世纪后期通过示范工程所传递的独立、分散式及因地制宜的水管理理念在 21 世纪初已经被付诸日常水务管理实践。同时人们对水的认识也大有改变，水不再仅仅是用完即扔的物质，而是用起来十分珍稀的宝贵物品，就好像这些水是从后代人那里暂时借过来用似的。水的真正价值得到人们普遍认可，并借助市场及媒体的国际化将这一思想迅速传递开来。20 世纪遗留的水问题慢慢地逐一得以解决。

　　在 21 世纪的第 2～3 个 10 年里，在世界各地的城市里已经建立起了很多以水为纽带相互关联的社区，节水及保护水资源已成为家庭生活常识。城市及区域性的私有给排水机构都被纳入流域水管理规划当中，向各社区供水并对污水进行处理以供后续利用。慢慢地，水交易网络在社区间、城市间及与周围农业用地之间得以建立，管理自然水资源的政府水务机构与以盈利为目的的给排水产业之间的权责也分配清晰。

　　在 21 世纪的第 3 个 10 年里，水交易已经成为日常交易的一项内容，水价依据其真实价值进行征收。国家级水务机构仅行使监督及控制权力，并负责国际间及全球范围的法律法规协调工作。那些由独立的供水商及本地用水社区组成的全球化网络功能日趋规范，各用水社区之间相互协助共同解决本地的水问题，市民们积极参与水务及周围环境管理。城市栖息地得到全面改善，与水相关文化传统重新焕发生机。自然又重新回到城市里来，为人们提供物质甚至精神的庇护所。事实证明分质供水系统行之有效，并促进城市的可持续发展。

　　所有这些取得的成果都要归功于人类在世纪之交所获得的对水务问题的深刻洞察及对未来的准确预见。到 21 世纪下半叶时，整个世界将会形成一张反应迅捷的城市网络，稳定运行的概率也更大。恐怕到时人们早已忘记这些在 20 世纪末进行示范项目建设的先行者了吧。

　　本文是在 Geiger 教授以前两篇作品（1993，2001）的基础上整合而成的。

参考文献

[1] Bates，B. C.，Z. W. Kundzewicz，S. Wu and J. P. Palutikof，Eds.，Climate Change and Water.Technical Paper of the Intergovernmental Panel on Climate Change，IPCC Secretariat，Geneva，2008：210.

[2] Geiger，W. F.. Urban Drainage - A Technical，Environmental or Social Problem?[C] Proceedings of the International Conference of Aquapolises，Shanghai，November 17-19，1993.

[3] Geiger，W. F.. Global denken und lokal handeln.[C] WaterScapes – Planen，Bauen und Gestalten mit Wasser，2001.

可持续水管理
——从目标到行动

W. F. 盖格

杜伊斯堡—埃森大学，德国

1 引言

世界上许多国家，包括中国在内，都陷入了工业高速发展、城市人口快速增长与城市生活水平的提高及绿地灌溉等对水资源的需求量日益增大的困境。虽然从1900年至1995年，世界人口只翻了一番，但需水量却增长了6倍之多。为满足巨大的需水量，水资源被大量开采，无视土壤退化及对地下水的影响，导致水资源压力逐渐增大，出现了许多几十年前未曾预见到的新问题。渐渐地，人们也开始意识到要在21世纪内维持高速的发展，必须要对传统的水资源管理方法做出重大改变。

更确切点讲，传统的水资源管理在长期和短期的规划管理方面存在较大差异。长期规划（规划期长达10年）通常通过监测对现状进行评估，对流域内的水量及水质进行评价，提出具体的发展和保护策略。而短期措施（通常规划期短于1年）由于城市及工业的快速发展变化，需在短时间内为决策服务，随即付诸行动。此外，水管理方面的相关法律法规也滞后于新技术的发展，因而很难去判断某些快速的决策是否能够很好地融合到水管理的长期规划中去。若不能，则会导致高投资、低效益的状况出现。到目前为止，尚无一套统一的将长期规划和短期行动统一起来的水资源管理程序可供参照。因而急需一种可以对问题区域的现状进行快速充分评价，并且对各项水管理措施，不管是技术性的还是非技术性的，从投资及效益方面进行总体评价的方法。最后在技术方面，还缺乏将水管理与可持续性整合起来的有效技术措施。

未来的水管理还需特别注意由城市群向超大城市的发展演变，如现有几大城市集群，位于香港、深圳和广州之间的珠江三角洲，雅加达—泗水走廊（印度尼西亚）和日本位于东京、长屋和大阪之间的东海道走廊。据估计，到2025年，超过60%的世界人口将生活在城市内，90%的城市人口生活在超大城市内。

2 可持续性的原则及水管理目标

可持续水管理旨在保护水资源，并确保我们的后代仍拥有充足的水资源。然而由于人口和经济的快速增长，尤其是自然的水文及地理条件欠理想的地区，很难找出一种既能满足社会—经济发展需求又不会破坏水资源，也不会对后代的用水造成压力的水管理方法。为探索一种水管理综合途径，一些国际性组织，如联合国教科文组织早在20世纪70年代就通过其下属国际水文计划（IHP）将气象学家、水文学家、生物学家、社会学家、经济学家、城市规划人员及其他相关人员召集到一起，以期促进对水管理所涉及的各个方面的理解，并向发展中国家传播相关知识（如UNESCO，1987；UN，1994）。

2.1 全局性途径——必要的前提

可持续水管理认同水系统的复杂性及系统内各要素之间的错综关系。作为一种全局性的方法，可持续水管理将各方平等地参与进来，包括当地及区域性的权威机构、科学家与普通雇员、环保主义者与决策者、各政党政客、不管是执政党还是在野党，尤其是受其直接影响的人群。可持续性的发展定义为"满足当前人类需要但又不损害后代人利益"的发展（世界环境与发展委员会，1987-"Brundtland report"）。它还保障后代人能够获取同等于现代人或者更高的生活水平。换句话说，可持续发展就是要避免决策者们为他们今天所做的决策后悔。更进一步说，可持续发展意味着一个社区的生产和消费不应对其他社区的生态、社会及经济基础造成影响，从而影响其生活质量的提高。最后，可持续发展还意味着保护生物多样性，采取各种预防性措施。

由水资源的自然属性决定，水资源管理是一项跨部门的工作。因而当前社会的分工及政府机构的多部门设置成了综合水管理的一个障碍。如何加强各部门之间的相互合作和各学科之间的相互融合，成为可持续水管理实施过程中的最大挑战。而当前的不注重各学科知识之间的融合及采用从全局出发的方法来解决资源问题的教育体制也更加大了可持续水管理的实施难度。因而，要实施水资源的可持续综合管理，亟须行政体制的创新和从全局出发的教育改革。

从环境和生态角度来看，水资源管理的最重要目标便是保护生态完整性。就此而言，时空因素非常重要。一个区域内任何时期的物种数量取决于过去的基础，因而水管理仅关注现状是不够的，还必须关注历史因素和当前微生物所呈现的特征。

在制定水管理的社会—经济目标时，还必须考虑文化、地理、经济和政治状况。社会—经济目标与供给—需求、防洪及保护自然环境密切相关。毫无疑问，维持和保护人类生命财产安全无疑是水管理的最重要社会—经济目标。但同时在部分乡村和城市区域保持生态系统的多样性也是一项重要的社会—经济目标。许多发达地区的自然环境曾遭受严重破坏，而如今又必须投入大量的财力采取补救措施。正如一句谚语所说："预防强于医治"。

用钱来量化城市水管理的社会—经济因素并不能体现出水管理的重要性。这种分析应该对价格和价值加以区分，价格是当前的、短暂的因素，不能精确地体现物品的最终价值。通过采用货币来衡量价值的方法应囊括环境成本这一概念，它既涵盖了下面的环境效应，也包括某些措施可能对环境造成的破坏。

2.2 水管理的目标

安全、可靠、平等的供水是城市水管理的最重要目标之一。由于城市区的人口密度大，经济活动频率高，因而预防洪灾危害，尤其是保障人身安全，是城市水管理的第一要务。

卫生及保护地表水水质也非常重要，不管是从环境保护角度来看，还是从公众健康角度来说。很显然，水管理要同时满足各种不同的，甚至相互矛盾的目标以实现经济、社会及环境的协调发展。

水资源可持续管理的不同组成部分在不同时期的重要性并不相同。在社会发展的不同阶段，不同目标的重要性也有所不同。在前工业化社会，水管理的重点主要是供水、运输和灌溉。工业化时期，水管理的重点包括水力发电和废弃物的处理和运输。后工业化时期的水管理重点开始转向美学和生态功能。以水为基础的娱乐活动在工业化和后工业化时代都很盛行。图1描述了随着工业化的推进，水管理的重点变化及同时期的环境状况。然而，我们还必须认识到尽管各时期水管理的重点不同，但由于水和各种用途之间的相互依赖性，长期来看，对那些非重点的目标也不可忽视。

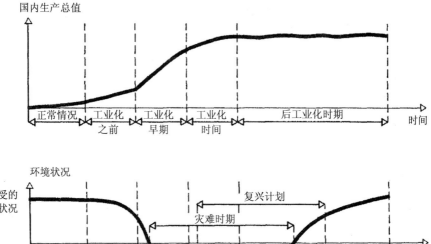

图1　工业化、经济、环境条件的相互作用（UNESCO，1995）

2.3　水管理中时空尺度的重要性

可持续发展不仅是多学科的融合，也是跨尺度的探索，既包括短期规划，也包括长期规划；既强调局部问题，也涵盖区域问题，甚至可扩展到全球视角。

在水管理中，对水资源现状的评价需要首先划定系统的边界。因为边界的划定视所需解决的问题变化而变化，因而边界的划定并非易事。举个例子，若要削减城市下游近郊水体中氨的毒性或减少耗氧量，只需对城区内的污染物浓度进行限制，并对其现状进行评价即可。而若要对城市下游远郊的水体中的硝化及沉淀过程进行研究，则不仅城区，整个污染源区内的农业输入也应一并考虑，如图2所示。此外，第一种情况下的削减措施可只考虑城区内的措施源，然而对于第二种情况而言，则要考虑各种用地方式下的所有可能发生的相互作用。

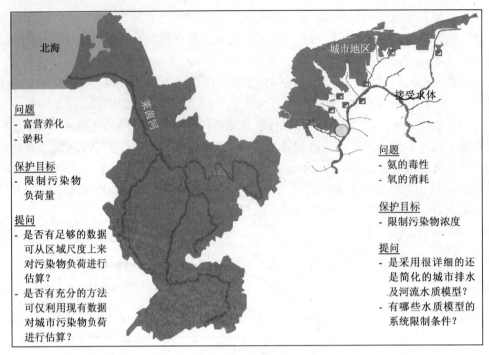

图 2 不同地表水体的保护目标及相关问题（Nafo and Geiger，2004a）

在后期对规划管理措施的影响评价中，对第一种情况可采用详细的物理模型，包括降雨—径流模型及污染物运移过程等，而对第二种情况，则只需采用一些相对于物理模型而言对数据要求较低的经验方法即可。如图 3 所示，水管理及物质平衡中的流域物理尺度、规划期限跨度及所采用的技术之间关系非常紧密。通常数据的丰富程度也是水管理中一个极大的制约因素。此外，若流域水文单元跨越不同国界，也应以整个流域为单元，采用相同的分析步骤进行研究。

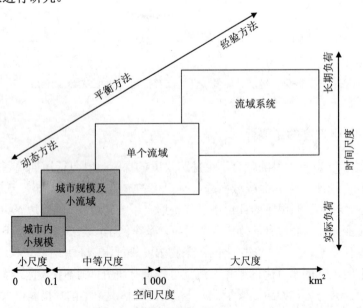

图 3 水管理的系统规模及时间跨度（Nafo，2004）

3 灵活的水管理决策支持系统

欧盟水框架指令为欧洲的水管理提供了一个集环境、社会和经济目标于一体的框架，但它并没有明确给出如何实现这些目标的固定程序，因而需要探索一套灵活的程序，以适应各区域的具体需求，并快速找出问题的源头，提出经济有效的解决措施。

3.1 欧盟水管理框架指令

水资源一直是欧洲法律所关注的重要内容之一，到目前共有 25 部法律法规，内容涉及地表水、渔业用水、洗浴用水及饮用水等，另外也有对不同排放限值的明确规定。新出台的欧盟水框架指令（EU，2000）提出了各种水体不管是量还是质，包括地表水和地下水，要达到一个良好的状态目标，并确定要达到这样一种良好状态的最终期限。此外，它还声明水管理应以流域为基本单元，并将控制排放与污染物达标排放相结合。

欧盟水框架指令所规划的环境目标包括避免当前地表水的生态状况进一步恶化，到 2010 年所有的地表水体及地下水都要达到一种良好的生态状况。在社会—经济方面，对所有用途的水价计收应能反映其真实成本，并要求在水管理决策过程中要尽最大可能地将公众参与进去。鉴于欧洲各区内自然环境千差万别，该指令并没有对水体的生态质量给出统一的标准，仅描述为"良好的生态状况"，因而水体的生物群落应与其天然状态相差越小越好。执行过程中，第一步是要对水体的现状生态质量进行评价，并与天然参照进行对比，得出水体的物理—化学及水文—形态学相关指标值。

每个流域都应开发各种监测项目及对水体状态进行定期观测，对地表水需进行生态及化学测量，地下水需监测化学指标及水量变化。此外，还需对各流域内的所有用水进行经济分析，包括取水和配水、污水回收及处理、相应的水价、各经济部门之间的成本分配及对供需水进行长期评价。

在欧盟水框架指令实施的过程中，某些特例可能会在将来引起政客们的关注。另外还需要将高度城市化、工业化地区及农业和畜牧业比较发达地区内的常规改变和剧烈改变区分开来。这样一来，对于那些需要付出较高的代价而能获得生态、社会及经济效益高的修复，就可将其要达到良好状态的最终期限（2010）适当延后。另外，在所定的目标太高，根本无法达到或是要达到这个目标的成本太过高昂的情况下，还可将所定的目标适当降低，如德国西北部爱莫希尔（Emscher）流域的情况。

3.2 实现可持续水管理的步骤

鉴于可持续水管理的确切目标及实施程序很难确定，各地区内的自然地理及水资源状况差异也较大，在对特定破坏进行修复的成本不能被社会普遍接受的情况下，需要寻求一种灵活的程序来对目标进行适当的调整。图 4 中给出了水管理的步骤顺序。

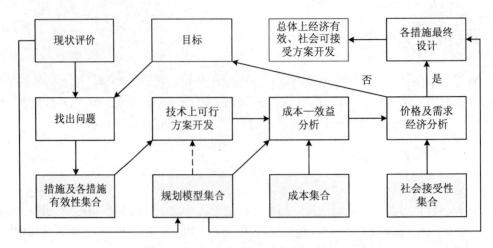

图 4 水管理的步骤顺序

水管理工作始于对现状水文、地质、环境、社会—经济及相关状况的评价,再将评价结果与相应目标进行对比。这些目标可能只是一些大致的目标,如减少洪灾的危害或改善市容,或者可以通过法律法规进行量化,如确定所要达到的水质级别或限定排污浓度。以往的现状评价需进行大量的监测及调查工作,现在许多水质问题都可以利用卫星数据进行分析,从而可以节省大量时间。

将卫星数据分析与过去对农业、畜牧业、城市用地等污染源观察的水质恶化目录相结合,便可很快找出流域内存在的水问题。通常问题越严重,越容易被发现。此外,监测项目还应有利于利用所得数据对长期规划的各种数学模型进行校正。

水管理规划的下一阶段便是寻找解决问题的最为经济有效的措施集合。当前的一种趋势是利用各种模型来向人们定性及定量地展示各种措施的效果,然而人们常常忽略一点,所有的模型都有其局限性,且其模拟结果的好坏直接取决于输入数据的多少及精确程度。

有些时候从模型得到的结果甚至会掩盖真相。因而这里推荐首先将各种措施进行收集,对其实施效益及操作成本进行估算,组成一个工具箱,其中既包括技术性措施,也包括非技术性措施。表 1 中给出了一个工具箱矩阵的例子,表 2 中给出了一些城市供水及排水方面的一些实例措施。以这种工具箱为基础,可以挑选出不同的措施组合开发出各种不同的水管理方案,最终再将这些方案从其环境及资源成本方面进行对比。

表 1 工具箱矩阵组织结构:以城市地区水管理措施为例

水管理措施	目标				
	节水	维持水量平衡	防洪	控制河道侵蚀	雨洪水污染防控
节水型用具	P, H, A	P, H, A			
屋顶雨水收集池	P, M, C	P, M, C			
中水回用	P, H, E	P, H, E			
污水回用	P, H, E	P, H, E			
绿色屋顶			P, H, E	P, H, E	
透水路面			PS, M, A	PS, M, A	

水管理措施	目标				
	节水	维持水量平衡	防洪	控制河道侵蚀	雨洪水污染防控
地表入渗			PS,H,C	PS,H,C	
渗透式浅草沟			PS,H,C	PS,H,C	
渗渠			PS,M,E	PS,M,E	
含水层储蓄和恢复系统	PS,H,E	PS,H,E			
地下水回灌		SD,M,E	SD,L,E	SD,L,E	
水土保持				D,H,A	D,H,A
过滤带					SD,H,C
响应式街道布局					SD,H,C
雨洪收集池					SD,H,C
水塘、湿地					SD,H,E
土壤过滤池					D,H,E
景观设计					

位置
P=小尺度
S=中等尺度
D=相对较大尺度

效率
L=低
M=中
H=高

成本
E=昂贵
A=平均
C=低

表2 最佳管理实践技术措施实例

生活节水技术	
	节水型水龙头 流速恒定,不随水压变化而变化 可直接安装进水龙头或淋浴头内
 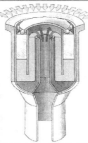	无水小便器及坐便器 表面涂层光滑,小便可全部排入虹吸器 小便器的表面覆盖可生物降解的消毒物质 依使用频率而定,表面涂层每年要更新多达4次
	中水回用可有效削减用水量 有无生物处理的系统均存在 供水安全级别较高 运行噪声低

家庭雨水收集	
	存储之前先要进行过滤 已存在自净型过滤器 新技术可使过滤效率达到90% 研究主要着眼于如何去除可溶性物质
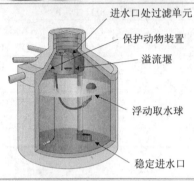	通过沉淀、浮动抽取球及溢流虹吸管进行处理 可将过滤器整合到储水罐中 每年仅需进行一次检查 使用5~10年后才需进行清理
分散式雨洪水下渗	
	透水地面从源头处削减径流 通过植被、入渗可对雨洪水进行很好的处理 维护简单方便 使用经过批准的产品可避免堵塞
	地表或地下土层贮存/入渗设施以增加地下水补给 来水必须不含悬浮或沉淀物质 利用地下蓄水以节省地表空间
集中式排水系统组件	
	集中式系统需要大型排水管道 因而推荐采用半集中及生态雨洪处理方法 去除可溶性物质

	依防洪标准不同，有必要采取集中式措施 需防洪空间较大 防洪设施内的空间可加以利用，但较危险 需防洪体积较大，1 000 m^3/hm^2
	在大城市内，若不是采用分散式污水处理系统，则需要建设很大的污水处理设施 大型污水处理厂的运行成本较高 运行不善可对水质造成很大影响

3.3 灵活的运行及控制程序以促进城市水管理

东南亚的城市发展异常迅速，很难对其城市发展进行很具体的预测，很难进行污水及雨洪水污染控制措施的长期规划。因而推荐设计一些短期（5 年）的污染控制基本措施，并且允许后期进行改进，其优点在于一旦发现这些设施的容量不足以满足需求，仍可通过采用届时的先进技术对其进行补充。

如图 5 所示，对于这些措施在城市地区的实际应用，应对现状进行更为具体的分析评价。在计算各设施的大小时，应充分估计其在规划期末的各种负荷，并根据现有导则进行设计。

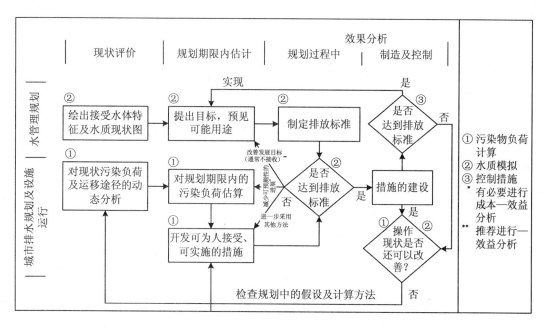

图 5　接受水体的排水及污染防控设计流程

下一步便要对规划期末的各种污染物排放情况进行估计，可采用手工计算或是数学模型进行模拟，并与现有排污标准进行对比（如果已经存在相应标准的话）。若排污标准不达标，则需对设计进行修改以达到排放标准。若该项标准仍不存在，则在该措施实施以后，需通过监测对其影响进行控制。若实施后效果不如预期，则应在初期运行后对其进行调整，若效果仍不理想，则应对其进行升级。整个规划、实施及控制流程如图 5 所示。

4 小流域内点源及面源氮负荷量评价

根据欧洲水框架指令（EU，2000）的要求，河流管理规划工作应以流域为基本单元，如大河流域，但其影响及各种管理措施的影响都应在中等尺度范围内进行调查，如一些小的流域单元。从流域氮的排放来看，不管是点源还是非点源排放，都应看做是人类活动对地表水体造成的压力，对其预防措施旨在避免对下流远距离的水体造成长期的影响，如沿海地区的富营养化，同时也避免造成对接受水体的急性氨氮毒性。下面将以德国境内的 Dhünn 流域为实例来分析小流域内点源和非点源污染氮素的排放，评价其对下游水生态系统及其近处接受水体的影响。为此，采用了不同的物质平衡方法对点源和非点源污染的氮素排放负荷进行了估计，并分析了非点源氮素的排放动态变化及其对上、下游接受水体的影响。

Dhünn 流域面积约为 82 km^2，其中城市占地约 16%，可耕地占 39%，主要的点源污染为境内两座城市的排水区，一个位于流域上游（3.3 km^2），而另一个位于流域下游（3.2 km^2）。

4.1 规划工具

本研究采用 MONERIS（Modelling Nutrient Emissions in River Systems）（Behrendt et al.，2000）集总模型对点源及非点源的氮排放负荷进行估算。因对非点源氮素的动态分析需采用时间离散模型，因而采用了 SWAT（Soil and Water Assessment Tool）（Arnold et al.，1990；Neitsch et al.，2002）对其进行分析。接受水体的急性氨氮毒性采用 BWK-M3（2001）法进行评价，其阈值设为 0.1 mg NH_3-N/L。

4.2 点源及非点源氮素排放负荷

图 6 中给出了对流域内通过各种不同运移路径及各不同分区内的氮素排放负荷的估计值。假设氮沉积量为 25 kg/($hm^2 \cdot a$)，且不考虑湖泊内的氮素负荷，则 1996—1999 年内 Dhünn 流域内总的氮排放量约为 270 t/a，所估算的氮排放负荷，包括湖泊的氮素排放及河道内的氮素损失，仅与观测值相关 2%。

面源污染（可耕地）被认为是向河流排放加营素的主要污染源，占总量的 79%。而点源污染仅占向河流排放氮素的 21%，其中污水处理厂排放量占 18%，城市管道溢流排放占 3%。尽管如此，一些分汇水区点源的氮排放量达到了 35%，包括 8% 的污水溢流。

基于以上数据，人们可能会产生一种错觉，认为点源污染，尤其是污水管溢流，对于下游水体的影响是可忽略不计的。其实，在对水体的评价过程中，污水管与污水处理厂之间的相互作用，以及氮的背景贡献量及土地利用方式的影响都应予以考虑。氮的背景值对总氮排放贡献量为 43 t/a，可耕地的氮排放量为 171 t/a，即为 54 kg/($hm^2 \cdot a$)。相比较之下，城市地区的氮排放量为 56 t/a（42 kg/（hm²·a）），也就是说城市用地占评价流域总面积的 16%，而氮排放量却占到 21%。同样计算，可耕地占评价流域总面积的 39%，总氮排放量却占到 63%，因而可以得出在当前研究中，氮的点源污染不可忽略不计，因而仍应采取相应措施对点源污染进行防治。

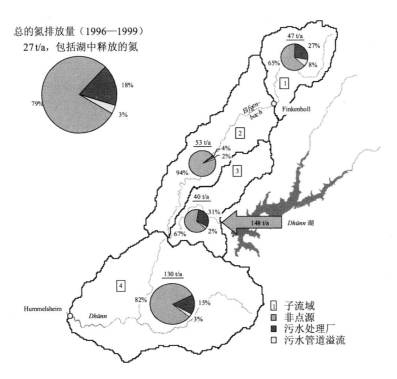

图 6　1996—1999 年研究区内总的氮素排放情况，假设氮沉积模数为 25 kg/（hm²·a）
（Nafo et Geiger，2004b）

4.3　接受水体的氨氮毒性

为对氨氮毒性进行评价，假设非点源负荷和背景氨氮浓度均为 0.3 mg/L，地表水的 pH 为 8，水温 20℃。根据影响矩阵判断，由于污水管道溢流及两座城市的污水处理厂排水，接受水体内的氨氮浓度全部超过 0.1 mg/L 的阈值（图 7，左）。氨氮毒性只有在地表水 pH 为 8.5 时才会出现。当污水溢流量较少时，由于地表径流与污水的混合性较差，此时的有毒氨浓度最高，计算结果如图 7（右）所示。仍假设上述非点源排放负荷及背景浓度为 0.3 mg/L，水温 20℃，则不管 pH 是 8 还是 8.5，河道内下游水体中氨的含量都超过 0.1 mg/L 的阈值。

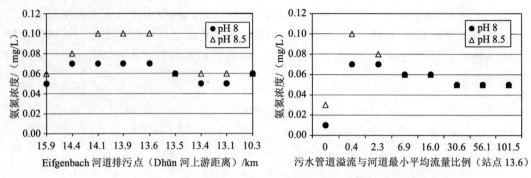

图7 各排水点（左）及 Dhünn 流域上游 13.6 站（右）氨氮浓度计算值
（Nafo et Geiger，2004b）

4.4 非点源氨氮负荷动态变化

使用物理模型，模型校准非常重要。本研究采用校准后的 SWAT 模型对非点源氨的排放动态进行了评价，结果表明其水量平衡及物质平衡效果良好。

为更好地评价非点源氨排放对河流的影响，本研究也对不同地点的浓度—历时关系进行了模拟，结果如图8所示。该图表明 Dhünn 河下游水中的氨浓度低于 Eifgenback-Dhünn 汇合处的氨浓度，由于 Dhünn 湖来水对河道内水流的稀释作用，也表明 Dhünn 湖对下游水体的水质有明显的影响。图 8 中的浓度—历时曲线还表明最大日平均氮仅出现在每年的干旱期内少数的几天，表明非点源氨的排放对河流的日平均最大氨浓度影响甚微。

若无 Eifgenback-Dhünn 湖汇合处湖泊来水的稀释作用，Dhünn 河下游水体中干旱期的氨浓度甚至是超过 0.3 mg/L，如图 8 所示。这表明若没有 Dhünn 湖的存在，流域内上游的点源污染源需认真对待。

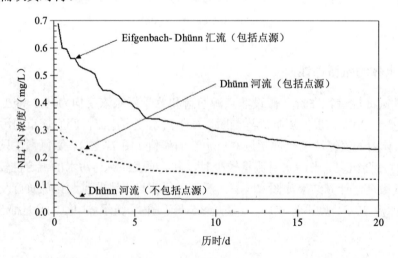

图8 1996—1999 年 Dhünn 河不同站点处氨氮浓度—历时模拟结果
（Nafo et Geiger，2004b）

5 结论

本文阐述了当前水资源管理实践存在的主要不足，描述了结合现有经验和最新的技术来加快设计过程的可能性。建议采取经济节省的措施。特别是大城市的水短缺、水污染、洪水问题证明了传统的供水、防洪和水污染防治策略已经达到了它们的极限。如今，我们已经有了新的策略，特别是在北京的示范项目中表明，综合最新的节水技术、就地储蓄雨水径流、处理和入渗可以极大地减轻水资源的压力。

当今，农业和城市化的不断发展需要采取补救措施。因此，重新审视传统的方法和采取新技术措施的时机已经来到了。然而，正如图1所表明的，只有在后工业化社会才能出现工业化、经济和环境状况的相互作用。如果能够有效地结合工业化国家的经验和当今的技术，而不只是拷贝过去的系统，我们将会更快地实现可持续性的水资源管理。

防洪与地下水回灌
——中德合作北京示范项目

W. F. 盖格

杜伊斯堡—埃森大学，德国

1 引言

城市供水安全与其周围水资源丰富程度密切相关。许多城市的供水主要依赖于当地地下水的开采、水库及河道径流。过量开采地下水会导致地下水位下降，若在沿海地区，当地下水位低于海平面时，还会引发海水入侵。

北京水资源极度缺乏，位列世界十大缺水城市之中。除缺水外，城市洪灾也是北京面临的威胁之一。北京年平均降水量 640 mm，而 80%集中在汛期的 3 个月内。大量硬化地面的铺设极大地减少了地下水的补给，使得汛期内高强度的降雨很快形成地表径流，增大了洪水发生的概率。在图 1 中将北京与德国埃森（代表中欧地区降水状况）一年内不同时期的降水量进行了对比。

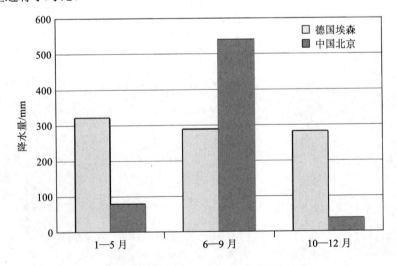

图 1　北京及德国埃森各季节降雨量对比图

北京地区按其城市开发利用程度可划分为 4 种类型，针对不同类型，需采取不同的防洪及地下水回灌措施：

- 类型1：由早期城市开发形成，并保持其利用方式不变
- 类型2：由早期城市开发形成，后期经过改建用地有所调整
- 类型3：重建、改建或城市利用加密区
- 类型4：由农业用地开发利用形成的城市

在本合作项目中，经过仔细勘察，最后选取了5个区域进行示范项目建设，囊括了上面区分的4种用地类型。本文将对其中的3个示范项目进行详细阐述，包括：北京市地质工程勘察院处降雨径流处理项目（类型1），新开发住宅区天秀花园内地下水回灌项目（类型2及类型4）及水利基础大队处节水设施示范项目（类型3）。

2 水文及水文地质条件对相关设计的重要性

对于任何水系统设计而言，在设计过程中，除土地利用方式和当地气候条件外，还必须对当地水文条件及水文地质状况进行评估。以与防洪及地下水回灌紧密相关的排水系统设计为例，任何粗略甚或错误的估计都有可能导致系统容量过大（浪费财力）或过小（增大洪水风险）。因而，必须对水文条件及水文地质状况进行仔细的评价。

2.1 水文条件及设计降雨推求

项目进程中共有7个气象站的水文数据可供利用。鉴于大多数示范项目都位于海淀气象站附近，因而设计过程采用了该站的数据。在进行设计降雨过程推求时采用了石景山站的资料，因该站资料时间序列较长（1981—1999年），且较为完整。石景山站最大降水出现在1994年，总降雨量达775.2mm，蒸发潜力为1946.9mm；1997年降水最少，仅333.0mm，蒸发潜力高达2154.5mm；多年平均降水量为553.0mm，平均蒸发皿测得蒸发量为1826.1mm。在该站19年降雨记录的基础上，推求得出降雨历时—强度关系及设计降雨过程，如图2所示，用于后期工程中雨水收集及存储设计的大小计算。

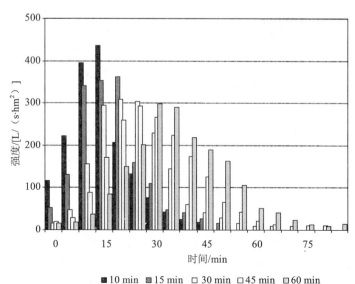

图2 确定不同部件尺寸的设计降雨过程图（Dorsch Consult in BMBF，2002）

2.2 水文地质及地质条件

在项目进行前，天秀花园内原有15个钻孔。为更好地对地层属性进行分析，又于2000年在天秀花园和水利基础大队另外打了18个钻孔。大多数钻孔的深度约8m，有一个更深一些。从这些钻孔来看，由于差异高达1.5m，很难确定砂和淤泥层组成的表层的厚度。此外，大多数钻孔集中在示范二区的中心地带，因而必须对测量结果进行外推。德方项目合作伙伴WASY公司利用TIN（不规则三角网）方法和ArcView-extension 3D-analyst软件对当地地层状况进行了模拟，以确定表层土壤的底部高程，得出了当地地形信息（图3）和地下砂层分布信息（图4）。

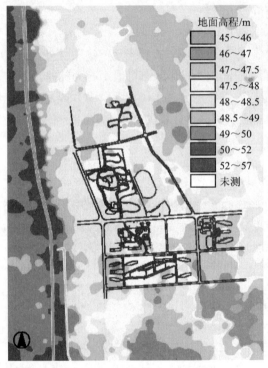 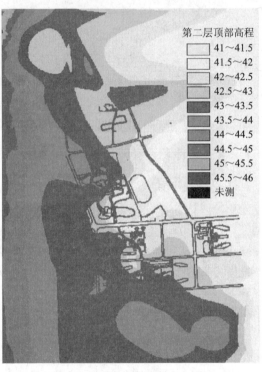

图3　天秀花园区数字地形模型　　　　　图4　模型第二剖面显示地下砂层分布信息
　　　（WASY in BMBF，2002）　　　　　　　　　（WASY in BMBF，2002）

天秀花园区土壤表层深度变化不一，不透水，下覆透水砂层，再下面为一半透水砂层，最下层为透水粗砂层。

为对地下水位情况进行分析，也采用了示范项目区周围的观测站数据。由于地下水相关信息稀少，德方项目合作伙伴WASY公司采用了FEFLOW模型（Diersch，1998），结合现有数据，对当地地下水相关信息进行了模拟和分析，得出了许多详细的地下水信息分析结果。

模拟结果表明，水利基础大队处的现有水文地质剖面信息以足够设计之用，而天秀花园处的地下水信息通过各种模拟技术也得到了补充。所有为模拟付出的努力在很大程度上减轻了设计结果的风险，所得出的地下水模型也为后期评价地下水补给的效果奠定了基础。

3 雨洪水处理系统示范

在整个合作项目中共修建了3处雨洪水处理设施,此处详细介绍的是水文地质勘察院内的设施。该系统与一停车场相连,汇水面积9 200 m²,雨水流入排水沟,上覆格栅。系统设计峰值进水量采用5年一遇5分钟历时降雨量,计算结果为120 L/s。设计目标为用雨洪水尽可能地入渗补给地下水,在遇5年一遇降雨情况下,不会有地表径流产生。蓄水池的水通过两个回灌井回灌地下,每口井最大入渗速度为 4 L/s,合起来 8 L/s。系统容量大小采用Dorsch咨询公司的HYDROCAD模型通过流体力学径流模拟计算得出。

该系统结构如图5所示,蓄水池为方形,长、宽分别为23.25 m和6.50 m(外部尺寸,包含墙的厚度)。正常运行情况下,水位变化约为 2.90 m,这部分水位变化对应的蓄水容积为 391 m³,相对应降雨量为 425 m³/hm² 或 42.5 mm。降雨期间收集系统内还可贮存部分雨水。

3.1 初期径流池

相对于后期较清洁的雨洪径流而言,降雨初期的径流污染非常严重,这部分严重污染的径流应通过初期径流池处理后,排入公共污水收集系统。此处初期径流池的容积为18 m³,约折合为整个9 200 m²汇水区面积上2 mm深的降雨。

每次降雨过后,需将池内污泥用泵清空。为保护水泵并避免大颗粒污染物进入公共污水收集系统,池内最底部的污泥最好保留在池内。这也是为什么每次降雨后必须对其进行检查,如有必要,可用人工或吸罐清除污泥。

3.2 格栅

初期径流池蓄满水后,水流流过一处堰坝,流过6个可移除静止格栅(不锈钢金属冲压材料,网眼大小8 mm,每个格栅大小为750 mm×750 mm),总面积为3.37 m²。开口设置在比永久水位稍微高一点的地方,这样干物体可以掉下来并防止收集的有机物质腐化,而一旦水位稍微上升,它们又会被浸入水中,以确保格栅表面负荷最小。如果出现堵塞,水会通过在舱顶稍下方的墙壁上的开口溢流到蓄水池内。格栅也可用作油污收集器。尽管如此,若有少部分油污流入下一室,也不会造成很大后果。

需要对格栅进行定期检查,如有必要,可对其进行人工清洗。若格栅上配有手柄,则更易进行拆除及清洗。

3.3 沉淀收集室

池内永久水位为 1 m。通过格栅后的雨水进入沉淀收集室,其中所携带颗粒污染物慢慢沉淀。水流通过6个弯头溢流管流入下一室。通过这一装置,雨水中剩下的油污可以被截留下来。在出流速度为8 L/s的情况下,此舱室的表面负荷仅为 q_A =0.60 m/h,可以确保很好的沉淀效果。该室也需要不定期地清空,包括其中污泥,可以通过管道将其中污泥引入初期径流池,最终排入公共污水管网。

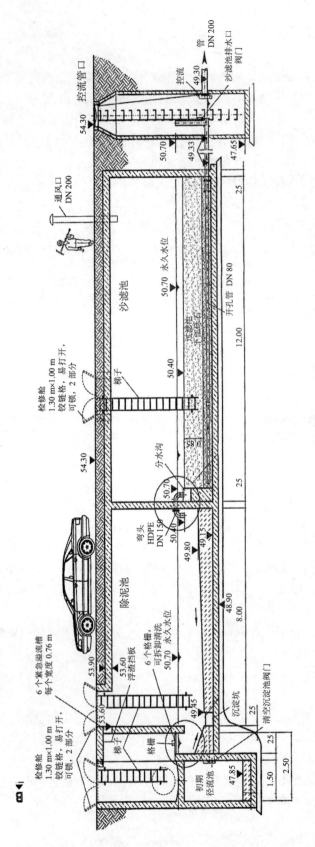

图 5 雨洪水处理系统平面图（UFT in BMBF, 2001a）

3.4 沙滤室

最后一个舱室内铺有 1.25 m 厚的沙子（过滤管道埋于 1.0 m 砂层下）。水从上部往下渗透，通过穿孔过滤管排出。当进水流速为 Q=8 L/s 时，室内沙滤的表面负荷为 q_A=0.40 m/h，为该类沙滤器的最大负荷。

沙滤分上下两层，粒径不同。上层质地较细，过滤性能较好，下层质地较粗，起支撑上层砂层的作用，并防止上层细砂进入过滤管道。根据下层砂层的特性，过滤管道选用不锈钢管，管径为 DN80。

沙滤室开口较大，以方便设施运行多年后更换滤料或进行其他维护操作。沙滤室要避光以防止藻类生长，因而开口处只可使用实心不透光的盖板。

3.5 水流控制检修孔

从沙滤室接出的过滤管道通向与回灌井相连的检修孔，检修孔内的水位保持恒定，底部较深。必要时也可通过活动潜水泵将该孔中的水排入回灌井。通常情况下，检修孔的排水潜力取决于回灌井的回灌速率。如有需要，也可在该处外加一个流量控制装置，通过测量孔内水位与回灌井内水位高差，便可对流量进行监测。

4 地下水回灌系统规划设计示范

在整个大项目框架下，每处示范项目内都包含有地下水回灌设施，此处仅举天秀花园内两处回灌设施中的一例。该系统设计目标包括收集屋顶、道路及庭院的雨水径流，然后进行入渗，以补充当地地下含水层回灌水量。屋面径流收集后临时存储在收集管道中，经处理后输送到回灌井。为滞蓄部分雨水，部分屋面径流收集后直接流入一蓄水池，用于补充区内人工湖用水蒸发损失。

4.1 蓄水池规格

蓄水池容积大小取决于区内水面蒸发速率，更重要的是干旱少雨天气持续时间。由于北京年内各月平均降雨量变化较大，在计算过程中，采取了假设不同蒸发速率值来推求所需蓄水容积的方法。假设湖面面积为 4 260 m²，图 6 中给出了推求蓄水池容积的图表解法。

举个例子，假设干旱期长 60 d，该时段内每月平均蒸发量为 80 mm，则可计算出蓄水池容积至少需要 682 m³。另外，旱期降水次数少，雨量也小，可假设复蓄水的概率不大。因而，即便建造一个非常大的蓄水池，也有可能出现旱期水不够用的情况。如果所建蓄水池太大，大部分雨水都用于补充湖面蒸发，则地下水补给量将会大大减少。

4.2 屋面排水系统构造

屋面雨水主要受灰尘和花粉污染，但程度较轻。此示范项目中，两处回灌设计各自配备一套处理系统，在屋面雨水流入回灌井之前进行轻度处理，其功能包括：
- 去除久旱初雨后的初期径流

- 去除灰尘，拦截漂浮物
- 向地下水回灌井输水

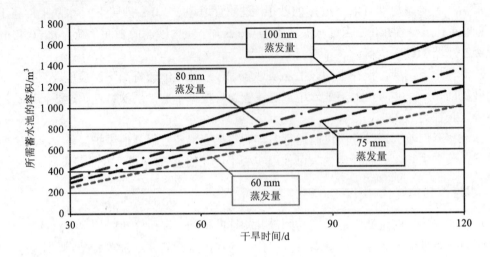

图 6 人工湖的蓄水池容积计算图解法

该系统的初期径流池可拦截每次降雨后的最初 2 mm 降雨。南北两处屋顶汇水面积分别为 1.17 hm²、1.12 hm²，初期径流池容积均为 21.8 m³。初期径流池蓄满水后，水流通过 DN 200 的管道流入回灌井。最后回灌井的水位达到与排水管内的水位齐平。回灌井的回灌能力受地下含水层导水率的限制。每处四口回灌井由 DN 200 管道相连，没有阀门。回灌设施及集水管道内的水入渗完后，再利用小型潜水泵（$Q=1\sim2$ L/s）对初期径流池进行清空。所用潜水泵必须能够抽取高度污染的污水和泥沙。为使尽可能多的水入渗地下，应尽量降低水泵的运行频率或缩短其工作时间。依所测得设施内部水位，水泵需能够实现自动控制。另外，需对整个系统进行定期检查。初期径流池内剩余污泥需进行人工或采用真空设备进行清除。

4.3 回灌井规格

回灌井建设的主要目的是确保大量地下水能够回灌地下，因而在天秀花园园区内南北设置了两处回灌设施，每处由 4 口井组成。前面提到的水文地质调查为回灌井的设计计算奠定了基础，包括地下水位、入渗系数、剖面高程及边界条件等。回灌井井径 2 m，上部为混凝土圈。区内 1998 年 9 月地下观测水位为有记录最高值，便采用了这一水位值用于后期的入渗速率计算。

由于入渗相关设计重现期不能事先确定，此处采取对多个方案进行模拟的方法进行重现期优选，对重现期 10 年、5 年、2 年及 1 年等方案模拟的结果表明井内相对应水位分别为 47.5 m、47 m、46 m、45 m 及 44.5 m（平均海拔）。从模拟结果来看，基于萧家河处 11 年地下水数据，6—9 月的地下水，十年一遇重现期下，回灌井入渗能力分别为 5.7 L/s（南部）和 4.7 L/s（北部）。五年一遇重现期下，入渗能力达到 6.7 L/s（南部）和 5.4 L/s（北部），上述数据均基于井内水位高出平均海拔 47.5 m 的基础上得出的。图 7 中给出了利用 FEFLOW 模型模拟出的在地下水不受扰动情况下在入渗后期地下水流场（南部回灌井，半天后）。

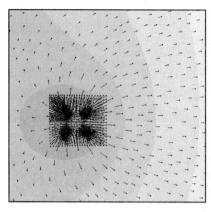 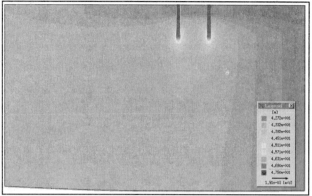

图 7　入渗过程中井下方地下水流向（左图）；入渗过程中东西剖面水头状况（右图）
(WASY in BMBF, 2002)

为减少相邻回灌井之间相互干扰，各处回灌设施内相邻两口回灌井的间距（中心之间距离）至少 17.5 m，否则会降低入渗速率。井与周围建筑物之间的距离一般为挖方深度的 1.5 倍，即距最近建筑物 12 m。回灌井上部护壁采用混凝土管，高度 0.5 m 或 1 m。南部回灌井的深度为 9.55 m，北部回灌井的深度为 7.75 m。

入渗主要发生在竖井下部垂直方向。由于挖方回填采用非黏性渗透性材料，因而通过各处设施内连接 4 口井之间的穿孔管（DN 200）也会有水入渗地下，这部分入渗水量并没有加入到模型中去。竖井底部填有约 1 m 厚粗砾层，上覆抛石以防侵蚀。此砾石层的入渗系数不得小于 5×10^{-3} m/s。如有必要，可对这一层进行更换。

5　节约生活用水的作用

节水的第一要务是让公众意识到节水的重要性和水的真实价值。此处节水示范技术方面包括在两幢住宅楼内安装雨水收集和中水回用设施。在降水稀少但人口密度较高的半干旱地区，如北京，将受轻度污染的家庭生活污水（中水）用于冲厕可有效地减少饮用水用量。另外，此示范项目的目的在于向人们展示，中水回用及雨水收集利用可很好地结合到家庭生活节水中去。

5.1　中水回用站

所选示范楼为新建居民楼，位于水利工程大队处，每栋 6 层，每层 10 户。图 8 给出了两栋楼内雨水、中水及黑水管线布置平面图。中水回用装置位于两栋楼之间。

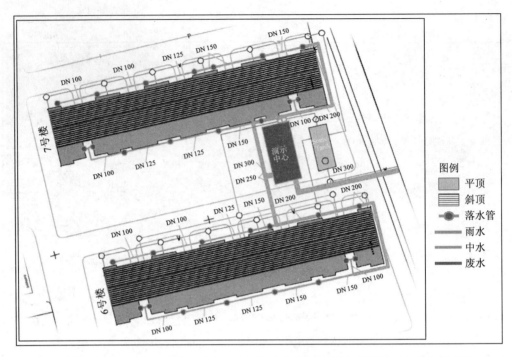

图 8　水利工程大队两处示范楼内雨水、中水及黑水管线布置图（GEP in BMBF，2001c）

为展示日常生活中在不影响舒适度的前提下进行节水的可能性，在 6 号楼内安装了节水型用水器具。作为对照，7 号楼内安装普通用水器具。但两栋楼内的抽水马桶均为 6L 型。

两栋示范楼顶的雨水经收集、过滤后存储在中水示范中心地下。6 号楼内的洗浴用水及面池排水经收集处理后回用于冲厕。生物处理过程中产生的污泥导入化粪池，同时该化粪池也处理这两栋楼里的黑水。雨水蓄水池的溢流雨水排入下水道。整个系统布置如图 9 所示。过滤设备使用合作伙伴 GEP 生产的开槽过滤器。

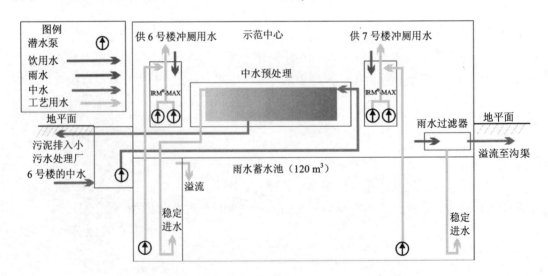

图 9　水利工程大队雨水收集及中水回用装置布置图

雨水收集及中水系统都配有调控设备,测量水流通量及池内蓄水量多少。遥控监测和维护已经成功运行多年。

5.2 设施规格计算

通常,蓄水池容积大小取决于年降水量多少,在北京还必须考虑如图1所示的降水分布极其不均的影响,另外平坦屋顶和倾斜屋顶各自所占比例,最终计算得出屋面排水系统可收集水量为 $668\,m^3$,仅占冲厕用水量的 29%。

在海淀气象站 1981—1993 年日降雨数据的基础上,通过模拟发现若设计蓄水池容积 $120\,m^3$,90%的屋面径流能够加以利用,且 14 年内仅发生 30 次溢流。然而,在 10 月中旬到翌年 4 月中旬期间,蓄水池几乎是空的。

中水回用站设计能力为 10 000 L/d,然而计算表明冲厕需水量仅为 6 000 L/d。从图10中给出了水利基础大队处的中水回用站基本构造。

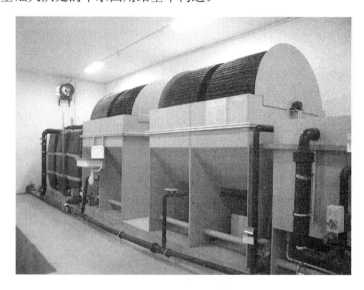

图 10 水利基础大队处中水回用装置

中水由于水量较稳定,随季节变化不大,因而在半干旱条件下中水回用是一项很适用的节水措施。然而与雨水利用相比较,其涉及技术更为复杂,投资及操作成本也较高。

6 结语

在半干旱地区进行与水相关设计时,应考虑充分利用雨水收集措施,收集尽可能多的雨水,不管是用于地下水回灌、家庭生活用水,还是商业性利用,并考虑与污染较轻的中水一起利用。从长远来看,这些方法具有很大的节水潜力。必须检查是否能在大面积的城市区域使用分散和独立的系统,例如堆肥马桶、ECOSAN 系统(LANGE 和 OTTERPOHL,2000)等。尽管如此,这些系统看来适合人口大约为 50 000 的郊区。

在进行系统设计时,需要对当地的水文及地质数据进行详细的分析,以尽可能地避免操作风险和错误的投资。在人口密度较高的城市进行水相关设施设计时,建议采用先进的规划设计工具，如 HYDROCAD 和 FEFLOW。虽然这些工具要求较高，强迫用户对当地情况进行详细分析，但最终这些分析将有利于帮助用户找到更优的解决方案。

最后不得不提的是，在各种节水措施的操作过程中，公共意识起到非常重要的作用，并能进一步加深人们对水的真实成本的认识。

从工程角度看成本效益

W.F. 盖格　P. 梅耶
杜伊斯堡—埃森大学，德国

1　制定成本效益措施的必要性

对水体造成巨大影响的因素有：个人、社区和工业以及包括农业径流和采矿在内的扩散源。由于诸如温度上升、溶解氧缺乏、富营养化等特别的水质问题，水的状态总是不能达到最好。在不同层面上需要制定并实施一些措施——流域层面和水体层面。层面不同，缺失程度不相同，规划条件和规划重点也不同。

对有效实施成本效益措施而言，建立对有效措施及其效益和成本的正确认识是非常必要的。措施制定要尽可能兼顾地域特点、地域条件。因此，最终的措施可能与总的规划有所不同。为了尽量减少这些不同，有必要开发出一套适用于所有情况的最终方案。

这一点，对于作为《欧洲水框架指令》河流流域管理运行方案的"欧洲水策略"核心资料来说，尤为重要。作为重要的文本资料，这些运行方案不仅需要列出保持水体现状的保护性措施和改进措施，而且还履行着确保成本效益措施及其维持措施得以实施的经济职能。

2　取得成本效益的步骤

针对流域内的水缺乏，需要制定出一个界定成本效益的客观的措施。最可行的措施是可以在流域内的水源本身实施的措施。实施地的环境应该允许这些措施的进行并确保对水体/流域没有不良影响。为此，把所有可行的措施列出目录以供选择是非常实用的方法。当然，最可行的措施是停止水缺乏现象。目录（工具箱）应该具有总体上的可行性，同时，适用于各种工程。在这个工具箱中，不同的措施要根据效益和成本进行不同的阐述。按照图1中的建议的步骤，能够取得成本效益。

受德国北威州环境保护、农业和消费局委托，杜伊斯堡—埃森大学、魏玛州的包豪斯大学、波鸿的德国创新与结构调整财政政策研究所，联合负责开发并试验上述提及的方法，以便制定出成本效益措施。

为了便于理解，我们先来讨论这些措施的效益界定的问题，然后对不同的成本进行界定。最后，讨论成本效益措施的组合利用及价值评估问题。以上方面的问题，要在保持水体的良好生态和化学状态的基础上进行讨论。

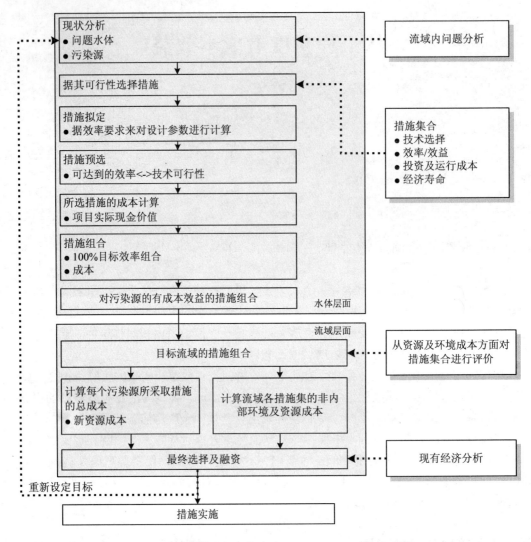

图 1 取得成本效益的工作步骤（LONDONG et al., 2005）

2.1 （治理）措施效益的定义

在特定的实际应用中，根据水体的生态和化学现状、目标水源地的水流集中度不同，治理措施的基本效益也不尽相同。因此，需要对目标水源加以辨别和划分，进而从图 2 中选择可行的措施。然后，再根据当地的情况检验所选措施的效益。建议按照 10%的梯度来阐述这些措施的效益；对水体要求条件与水体实际条件的不同进行 100%的阐述。最后，这些措施的效益进行累加，以使目标水源治理措施的效益达到 100%。根据目录中所列出的方法，可行措施的效益应该可以计算出数值。例如，假设水体的目标生化需氧量要达到 6 mg/L，目标水源的生化需氧量为 18 mg/L，出水流量为入水流量的 1/3 时，生化需氧浓度为 3 mg/L，出水生化需氧量为 18 mg/L 的时候，那么，该水体中平均生化需氧量浓度为 9 mg/L。由此可见，采用降低来水生化需氧量从 18 mg/L 到 9 mg/L 的方法，能够使成本效益达到100%。

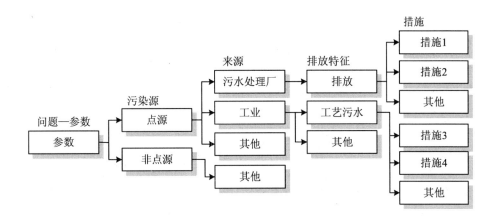

图2 根据水源识别治理措施的总体步骤

2.2 成本的定义

衡量不同治理措施成本的指标包括：投资运作成本和附加成本（附加成本是指环境资源成本）。投资成本是指建立治理措施的成本。运作成本，指该治理措施存续期内的运行和维护成本。环境成本的定义为"用水过程中或其他环境应用过程（包括水生态系统生态品质的降低、土壤盐化及土壤生产能力的降低）对环境和生态系统所造成的损害的成本。资源成本是指因超过资源自然再生能力而为其他资源的使用者带来可预见的机会成本，例如地下水超采。投资成本和运作成本是私人化的成本，而环境资源成本既是私人化的成本，也是公共成本。

为了使水的状况得到改善，要根据成本比较方法对选定的治理措施在措施的执行期内进行投资及运作成本的比较。每一年内，物价的上涨和下降因素都应该作为计算实际现金价值的因素加以考虑。在这种情况下，不同治理措施的成本才能够得以相互比较。附加的环境资源成本也应该加入成本分析中。

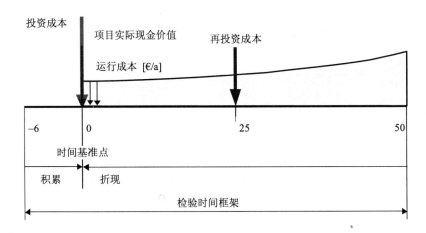

图3 计算不同治理措施工程的实际现金价值（LONDONG et al., 2005）

2.3 成本效益的计算

依据以上方法计算了不同治理措施的成本和效益后,这些措施要相互结合,以期达到成本最小化和效益最大化。首先,应确定一个使各种治理措施的总体效益达到 100%的指标。然后,所有被选定治理措施的效益应该和其他备选措施的效益联系到一起。因此,下次确定的指标就是相对于被选定的措施组而言的了。如果这时效益不能达到 100%,要按照程序继续选择治理措施,直到水资源缺乏的目标效益达到 100%为止。对于单个具体治理措施而言,实际的现金价值是由自主效益决定的。把每个工程的实际现金价值相加,得到措施群的合计现金价值。在这方面,不同措施组会达到相同的效益,但同时会产生不同的成本。效益成本最高的措施或措施组是工程现金价值最小的措施。不同指标条件下、不同措施的成本和效益如何组合(图 4)。首先,从措施目录里选出第一个指标(指标 X)条件下的所有相关成本效益为 100%措施。然后,计算这一措施的第二个指标(指标 Y)条件下的效益。现在,以指标 Y 为基准,从目录里选择其他的措施,以达到效益 100%,进而再计算出相关的成本。

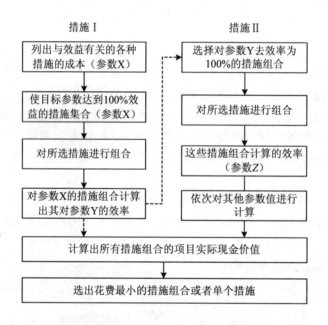

图 4　成本效益措施的价值计算 (GEIGER et al., 2005)

3　水体或流域治理考虑的因素

水体特征可以用各种指标来体现——水文指标、物理化学指标、生物指标和水力形态指标。下面我们将讨论是否水治理措施及其效果和相关成本要按照水体(河流流域)个体来识别或者是作为一个整体。

3.1 水体单独考虑的必要性

为了达到良好的生化状态,对每个水体进行分析是非常必要的。在分析过程中,所有对水体造成污染的污染源需要进行识别。在对污染源进行界定后,很有必要把所有污染源分为初级污染源和非初级(次级)污染源。初级污染源对主要流域(整个流域下游)造成影响,而次级污染源只对当地的水体造成污染。针对每种情况,要区分次级污染源的治理措施是否可以施行,初级污染源的治理措施是否足以减少污染。对于流入大河流的小沟渠而言,不去争取良好的生态和化学状况也不失为一种选择。所以说,对每一个水体进行分析是评估一个污染源是原始污染源还是次级污染源的重要手段。

3.2 河流流域规模界定

在任何一种情况下,都要对河流流域的规模进行分析。这关系着下游流域水体的入水量。对一个流域而言,只有根据 3.1 中的要求,做出详尽的分析后,才能决定是否采用所有的治理措施,或者是维持水体缺水的现状。换言之,在确定使用什么样的治理措施前,先要确定是否每一个单独水体要达到良好的生态和化学状态,或者说,对一些小型水体而言,达到整个流域范围内的生态、化学平均水平已经足够。

4 获得治理措施成本效益的范例

以下的应用案例在达到《欧洲水框架指令》的要求方面,适用于各个国家的环境管理部门。案例发生在利珀(Lippe)河流域(莱茵河右岸支流,流域面积 4 881.8 km^2),用于治理水温和氯化物问题。根据当地的实际情况,将治理水温和氯化物的治理方案被列出目录,然后计算出所选治理方案的成本和效益。表 1 是该目录的部分摘要。

表 1 备选治理方案目录摘要(LONDONG et al., 2005)

序号	方案名称	效率分析(EG)⇔参照标准	成本
5.2	利用热量:排水—热量转换器	Heat: $L_{Waermetauscher} = \dfrac{4186 \cdot 998 \cdot Q_E \cdot \Delta\vartheta_{W,Soll,Wi} \cdot ZEG_{Wi} \cdot \eta_{Waermetauscher}}{\approx 4.000 \text{ W/m}} \leq 300 \text{ m}$	投资成本——热量转换器,加热泵+其他部分成本、建设成本 IC≈35.126·e0.0151·L [€]无额外供水网络 IC≈128.777·e0.012·L [€]有额外供水网络 水泵运行成本,维护: OC≈298.4·L + 5 959 [€/a] 预期效益: OC$_{benefit}$≈-2.5 €/(a·m^2 相连接平面) 使用寿命: 水泵约为 10 年 热交换器约为 50 年

首先，所有针对水温和氯化物指标的治理方案是根据目标污染源的实际情况筛选出来的。图 5 对所有备选方案进行了大体介绍，可以就改善水温和氯化物指标进行有选择地应用。

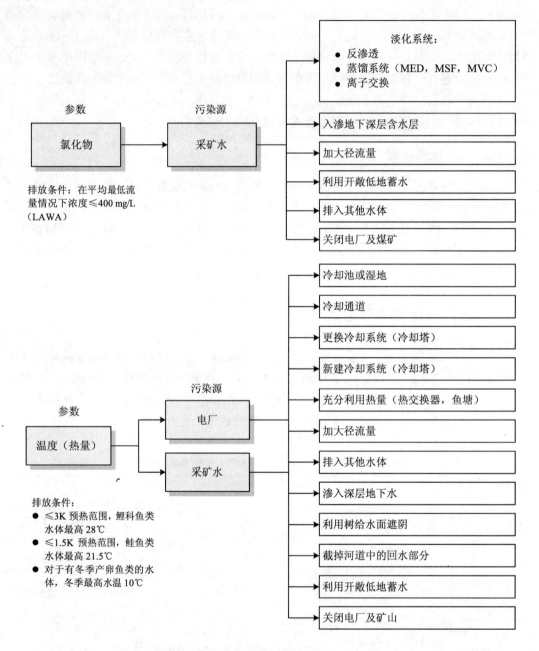

图 5　识别治理氯化物和温度问题的备选方案（LONDONG et al.，2005）

对于利珀河的水温和氯化物指标的问题，治理措施可根据图 1 的步骤进行。

4.1　问题评价与来源识别

针对利珀河的情况，有一个评估报告可供参考。根据评估报告，当地和该地区对氯化物浓度指标的要求为 400 mg/L，而在与瑞士接壤处莱茵河的氯化物浓度指标为不超过 200 mg/L（德国环保部标准，2004）（利珀河流入莱茵河，莱茵河是过境河流）。对于水温，当地和该地区及莱茵河入口的要求是夏季不超过 28℃（适于鲤鱼生存的水温，21.5℃适于大马哈鱼生存），冬季不超过 10℃。最大的水温升高幅度不能超过 3°K（鲤鱼生存水温）（1.5°K，大马哈鱼生存水温）（欧洲标准，1978）。

氯化物污染源可分为点污染源和扩散污染源两种。扩散性污染源来源于地球本身的条件，点污染源来源于煤炭开发、工业、污水处理和污水集中排放。煤矿污水、工业冷却水和工艺用水被排入利珀河及其支流。此外，还有冬天街道上的融雪剂通过雪水和雨水的盐分。污水处理厂接收的工业废水带有氯化物，另外，冬天的融雪剂里也含有氯化物。最后，流入河中的污水也带有融雪剂残留物。

造成水温问题的是包括煤炭开采，工业废水和污水处理废水在内的点污染源。煤炭开采排放出的含盐废水温度很高。工业排水主要是冷却水和热的工艺用水。图 6 对利珀河流域水温和氯化物指数有重大影响的污染源做了大致的介绍。

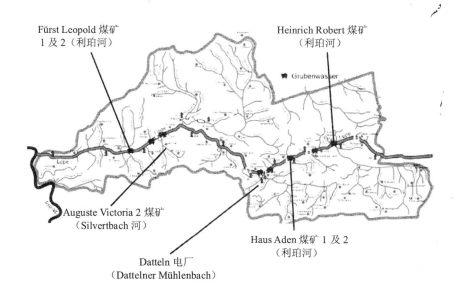

图 6　利珀河流域水温和氯化物污染源概览（LONDONG et al., 2005）

对问题流域的污染源进行分析以后，可以根据类似图 2 的方法选出针对点污染源的治理方法。然后，针对这些治理方法确定其成本和效益。

4.2　成本估算和成本—效益方法的选择

投资成本和运营成本可以进行精确的成本计算。环境成本和资源成本是很难独立计算的。所以，针对治理措施个体来说，只有投资成本和运营成本是可以估算的。

以下是针对奥古斯多·维多利亚 2 号煤矿污水排放的情况，根据上述方法对成本—效益措施进行的分析。表 2 是其效益及相关成本的估算。

表2 奥古斯多·维多利亚2号煤矿治理措施的效益及工程实际现金价值示例

(LONDONG et al., 2005)

措施集		效率	项目实际现金价值
1	将水引入莱茵河（39 km）	氯 100% 去除 温度 100%（SLC/WLC）	2 720 万欧元
2	将水引入莱茵河	（SLC）温度 90% （WLC）温度 80% 氯 80%	2 080 万欧元
	水塘	（SLC）温度 10% （WLC）温度 40% 氯 0%	20 万欧元
	反渗透	氯 20%	940 万欧元
		总计	3 040 万欧元

SLC—夏季负荷
WLC—冬季负荷

有几个治理措施组符合让利珀河流域水状况达到要求的条件。其中两个组合的投资成本和运营成本分别为1560万欧元和1670万欧元。但是，两个组合都存在向深层地下水渗透的问题。如果这两个组合不可行，下一个最经济的措施组合成本高达4520万欧元。因为具体细节问题，环境成本和资源成本是无法计算的。环境成本和资源成本可以用于比较措施组合，两者的成本同等重要。

赫尔姆·穆勒曾经就利珀河生态状况改进的付费意愿调查过一些家庭，并按照环境资源成本标准写出一份关于利珀河流域的报告。根据报告的效益转换率，利珀河流域治理后，每年可以产生1200万欧元的效益。当然，不同的治理方法会产生不同的环境和资源影响。这些对能源和水的影响大多数可以用市场价值来计算。因为某些数据缺失，只有对该地区规划改变的影响是难以估算的。

另一个问题是被选择的治理措施的经费问题。在上述提及的程序中，使成本最小化是共同的目标。但是，这些治理措施的费用要由污染的制造者来支付。因此，另外一条途径，也就是说，必须找到合理摊销治理成本的办法。

4.3 实践程序的修正

总结利珀河流域治理工程的结果，我们不难发现，可以用环境资源成本标准来阐述治理结果，但是，找到足够的与环境资源成本有关的数据却是不可能的。对于治理某些特殊指标的措施来说，环境资源成本的指标是空白的。只有考虑所有治理措施的整个程序，以达到被治理目标的良好生态和化学状况时，才能够大致按环境资源成本标准来阐述这一结果。所以，对单一的治理措施而言，只有对环境和资源的影响可以计算到成本效益措施中。在此基础上，一种新的达到目标要求的治理步骤被开发出来。在步骤6中，治理措施对环境和资源的影响可以估算到环境和资源成本中。

表3　理论治理步骤和为达到目标要求、根据《欧洲水框架指令》修正的治理步骤对比
（GEIGER et al., 2005）

步骤	理论步骤	实践步骤
1	对可能出现的问题进行分析和评估、识别污染源	对所有问题进行分析、对污染源进行识别
2	找出治理问题的措施	找出治理问题的措施
3	识别能让治理目标达到良好状态的措施	识别能让治理目标达到良好状态的措施
4	对可能的措施进行论证后作出选择	对治理措施的技术可行性进行论证
5	对所有治理措施和措施组合进行成本论证，包括环境资源成本	只在投资成本和运营成本的基础上对所有治理措施和措施组合进行成本论证
6	选择成本—效益治理措施	对措施组合进行预选
7	经费选择及治理措施（组合）的最后确定	对措施组合进行环境和资源成本评估
8	实施治理措施	经费及措施或措施组合的确定
9	治理措施的影响控制	措施的实施
10		治理措施的影响控制

5　总结及建议

到目前为止，选择成本—效益治理措施或措施组合的总的方法是靠专业人员的指示。这样，措施或措施组合的选择很大程度上就依靠专业知识的水平和治理目标。为了更加客观地对治理措施/措施组合进行选择，本文所述的具体选择程序为切实达到成本效益目标提供了可能性。

包括环境和资源成本在内的成本—效益程序不能为了包括环境及资源成本的单独的一项指标或治理措施而加以实施。河流水域总体质量的提高有助于减少环境及资源成本。有鉴于此，针对任何一个单独的指标或措施，都不可能计算出真实的环境及资源成本，环境和资源的影响也不可能作为评估一个措施程序的指标。当然，这里只是针对河流流域层面而言。

在对不同指标进行评估、对措施进行定义时，治理措施的效益成本程序只能有一个问题能够用手工方式计算；对于两个以上的指标，应该考虑用计算机进行计算。

当有必要应用复杂的质量模式时，就需要更加详细的数据。最后，成本计算方法会在必要的时候被指数系统代替。

环境保护措施的融资方式探讨

G. 乌兹伯格
德国施莱格工程咨询有限公司，慕尼黑

在过去的 20 多年里，中国的发展取得了令人瞩目的成就——经历了大范围的工业化发展进程，成功地解决了物资匮乏问题和文盲问题。在这一工业化发展进程中，以积极的目标——（人口增长、与国际市场的成功合作等）为核心是非常明智而全面的考虑。但是，由此而带来后果是：不可避免的一系列问题，例如某些资源的大幅减少以及对这些资源目前的保护与未来的合理利用问题。

世界上各种资源中最基本的资源——大气，近年来某种程度上变成了有计划治理的对象。根据《京都议定书》，国际排放率贸易已经开始生效。这意味着每个国家都要将其温室气体排放调整到自然排量标准，否则，就要向平均排放量低于标准的国家购买排放率。这表明，大气这一最基本的资源的利用是举世共同关注的话题。

水资源是继大气之后的第二大基本资源。对于水资源而言，同样有一个全球性的目标。在联合国《千年发展目标》中有一条是关于水资源的。这一目标是这样的：到 2015 年将无法持续获得安全饮用水和基本卫生设施的人口的比例将减半。

在中国部分欠发达的地区，改善水供应仍然是非常必要的问题。在工业化发展的过程中，新的问题也随之出现——水资源的消耗比水（数量和质量）再生的速度快。这一问题常见的后果是：地下水位下降、污染严重、病原体出现。例如在青岛周围地区，因为地下水位下降，地下水已经出现倒流现象。由此引发的后果是，海洋中的盐水渗透进地下水，会导致将来地下水无法饮用。

充足的洁净水对保证生活质量和身体健康而言是不可或缺的，对经济的进一步发展也是如此。特别是对于工业发达地区来说，以各种方式保护水资源更是迫在眉睫。

1 融资渠道

一般情况下，融资渠道有下列几种：①世界发展银行；②国家项目和基金；③股份制商业银行；④投资者——可以是本国的或国际的。

这样的排序是惯例排序，因为并不是所有的项目都能够通过最后提到的两种渠道融资。通过商业机构或投资者融资被认为是资本的倒退。所以，这两种融资渠道被认为是适合收费体制下的融资方式。

世界上有许多国际银行为包括饮用水和环境保护工程在内的基础设施建设提供资金

支持。本文中主要提到两个银行。第一个是最著名的并且是世界上最大的银行——世界银行；第二个是德国最重要的发展银行——德国国家发展银行，其缩写为"KfW"，意为"*复兴建设信贷公司*"。德国联邦经济合作发展部下拨的资金也通常来自德国国家发展银行。

2 世界银行

世界银行是联合国会员国（包括中国）主办的，为经济发展水平较低的国家提供服务，当然服务对象也包括发达国家和新兴发展国家。所谓新兴发展国家，是指经济正在发展进程中的国家。在经济发展方面，中国呈现出不均衡的状况，西北部地区仍处于贫困状态，南部和东部地区工业化程度很高，像上海这样的城市，可以称为新的经济发展中心。中国与世界银行已有多年的合作。

世行项目立项之前，需要制定一个《国家协作战略书》，它是国际开发协会和世界银行（又称国际复兴开发与发展银行）为贷款国制订协作计划的核心框架。《国家协作战略书》为世行合作国制定，上面指出了世行贷款协助能够最大限度地发挥扶贫减困作用的地区，并重点考虑合作国政府和主要利益相关者及其在合作国的作用和信誉。有关贫困的成因、贫困者的特点、体制发展状况、执行能力和执政能力也在战略书中有所体现。世行根据上述指标的评定结果、资金和技术协作的情况来进行贷款。这些协作战略是为世行加强与合作国之间的协作和合作而制定的。

世界银行提供资金支持的项目在政府层面进行，由合作国的财政部长与世行负责该国项目的经理进行接触。合作国的中央政府可以是项目发起者，省级政府或市级议院无权在世行直接发起项目。

通常，一个世行项目包含很多特定项目。当一个项目立项以后，该项目被分割成若干小型项目，然后再发布国际招标公告和进行任务划分。在世行项目子项目立项过程中，省级政府、市级政府、各协会绝对有施加影响的可能性。但是，对一些特定项目而言，任何决定都应该由中央政府和世行的项目负责经理做出。来自世行的资金既非无需偿还的拨款，也不是普通意义上的贷款。

世行的项目通常是大型和综合性的。因此，世行项目需要很长的周期。根据项目必须由合作国的中央政府根据《国家协作战略书》来完成项目的需要，项目前期准备工作可能要几年时间才能完成。如果一个项目已经存在，如果它是世行大项目的一部分就相对容易一些。

3 德国国家发展银行

在过去的 20 多年中，德国国家发展银行为中国提供了可观的资金支持。在大多数的年份内，中国不在从德国国家发展银行发放资金最多的 5 个国家之列。只有为数不多的年份，中国从德国发展银行获得的资金支持在 1 亿欧元到 3.5 亿欧元之间。在某些方面，德国国家发展银行十分重视与中国的合作，特别是污水排放和污水处理方面。

当污染非常严重的污水不能用普通方法处理必须使用高科技手段时，德国国家发展银行会提供合作服务。这些服务涉及规划、设备、人力资源培训等诸方面。服务得以推广的前提是，施行成本价费率。

有关流水线运作的专业知识转化日趋重要。以公私合作项目为例，中国的企业应当和富有经验的国际运营公司密切合作。

在融资和周期方面，德国国家发展银行的项目不如世行项目规模大。但是，项目前期准备工作要至少6个月。德国国家发展银行并非只与政府层面签订协议。

德国国家发展银行的融资工具：①复合融资；②促销贷款；③混合融资；④低息贷款。

3.1 复合融资

复合融资可用于基础设施领域内包括水供应、废水处理、环保技术等方面，符合发展政策、具备推广条件的项目。借款者可以是国家，也可以是得到国家批准的项目执行机构。

复合融资的期限根据项目的周期来确定。复合融资贷款既不是浮动利率也不是固定利率，其利率通常低于市场水平。根据合作国家的经济发展情况和项目本身的赢利能力，德国国家发展银行将采用复合融资的方式为合作国提供特制的融资服务。

复合融资可以是与货币供应相关联的并列贷款或无条件贷款。复合融资中必须有25%的捐赠款项作为官方发展援助金。

并列贷款是符合《经合组织共识》条款的非营利项目。换言之，并列贷款的款项适用于在商业融资条件下现金流不充足的项目。在这种情况下，捐赠款项的比例为35%。

3.2 促销贷款

促销贷款是为融资开发项目，主要是为无负债国家设计的融资工具，其条款与资本市场上的贷款条款相似。促销贷款使现行的金融工具更加完满，填补了发展贷款和商业融资制度的空白。促销贷款使德国的开发协作捐助工作更加完善，特别是对发展中国家的私营项目上变得日趋重要。在包括水供应等的基础设施建设方面，促销贷款融资可以为私人及公共项目提供贷款。

根据项目的开发信用情况，促销贷款是可以在任何一种情况下进行调整的。调整条件是以德国联邦政府的开发政策为基准。此外，德国国家开发银行要承受该项目的风险，借款人要有足够高的信用。促销贷款的发放条款与资本市场基本相同。

融资形式取决于借款人/项目的执行机构的信用及其特定需要。在这方面，德国国家发展银行可以提供灵活的贷款期限、货币种类、利率选择等服务，以满足合作方的需要。在基础设施建设领域，促销贷款的对象包括国家、私营企业或符合融资体制框架的项目等。在任何情况下，促销贷款都是与货币供应不相关联的。

特别是在经济基础设施建设项目融资方面，德国国家发展银行应尽快与合作方接洽，以便恰当采用复合的融资方式。由项目发起人/投资者提供的信息备忘录是初期风险评估的基础。至于详细的风险分析，要根据可行性研究报告来制定。最后，与项目参与者一起制定出一个包括抵押在内的可行的融资方式。

3.3 混合融资

在德国联邦政府资金储备预算的基础上，德国国家发展银行根据资本市场资金有利条件发放贷款，提高了德国融资合作的力度。通过混合融资的方式，根据对方的经济情况和项目本身的赢利能力与合作国之间合作，德国国家发展银行为合作国提供了特制的融资服务。

借款方可以是国家或者是由国家批准的项目执行机构。

在混合融资条件下，德国国家发展银行通过资本市场补充了预算资金。整个融资的下拨款项也符合官方发展援助条件的要求。项目的国家风险由德国国家发展银行承担，要求该项目由德国自由化出口保险（爱玛仕保险）承保（与货币供应相关联的固定金额融资）或者是由国外一级官方出口信用代理商担保（无条件混合贷款）。为此，混合贷款项目要符合合作双方各国的出口信用保险代理机构的条件要求。该项出口保险费用要在提供者的意向书中体现，并由项目执行机构承担。

混合融资的条件基本上根据向合作国提供预算资金的条件而制定。对于特别需要资金支持的、人均国民生产总值达到 875 美元的国家符合国际发展援助条件（年利率 0.75%，10 年宽限期，贷款期限 40 年）。其他所有发展中国家的混合贷款符合金融委员会条件（年利率 2%，10 年宽限期，贷款期限 30 年）。所有上述预算资金可以与假定的市场资金贷款相结合。

3.4 低息贷款

发展中国家也可以申请低息贷款。这样的贷款服务为更快的发展提供了额外的资金服务。根据合作国的经济发展情况和项目的赢利情况，通过这种融资服务，德国国家发展银行为合作国提供特制的融资服务。

不考虑贷款期限相对短的因素，低息贷款这一金融工具是专门为金融机构提供的贷款服务。原则上，这种贷款服务也可用于基础设施建设领域内符合开发政策的项目的推广，包括电信、能源开发和分配、运输业、环保技术和污水处理等方面。低息贷款不与德国货币供应相关联。德国国家发展银行只将来自资本市场的资金用于低息贷款服务。贷款的利息是通过联邦预算拨付的，所以，申请低息贷款的项目要符合官方发展援助的要求条件。拨款额度根据市场利率水平确定，且在合作协议签订的时候确定。

低息贷款的借款方可以是国家或者是国家批准的项目执行机构。

低息贷款的期限一般是 10~12 年，宽限期为 2~3 年。鉴于低息贷款的风险由德国国家银行承担，所以，款项只有在该银行认为能够承担项目风险的情况下才会发放。根据协议的条款，目前一般的低息贷款年利率为 4.25%，期限 10 年，宽限期 2 年，在年利率 5.00% 的情况下，期限为 12 年，宽限期为 3 年。

4 国家基金

在国际开发银行的融资手段难以满足需求的情况下，国家的努力是必不可少的。实际的国际援助可以解决大量的融资问题，但是最终的融资解决办法还是要靠贷款国自身的努力。

与水资源保护、分配以及防洪密切相关的工作可以分为两个基本类别。一种是完全属于公共利益的并必须属于公共支出范围的项目；另一种是可以由收费体制偿还款项的项目。

完全属于公共利益的项目基本上是行政机构用以组织水资源的利用与再生的法律或

法规范畴内的项目。有关费用和专业人才方面的工作可以通过国际援助来完成。但总体上，这项工作是国家本身的工作。这一点，也适用于防洪工作方面。

但是，在工业化达到一定程度以后，饮用水或工业工艺用水以及废水处理的费用可以全部由使用者和受益者支付。当然，这样的前提条件是收费制度的存在。据我所知，目前中国已有这样的例子。收费制度也是抑制浪费的有效途径。下面举例说明。

大约20年前，东德发生了政治变化，前德意志民主共和国加入了德意志联邦共和国。从那时起，市场经济体制被引入民主德国。那年仲夏，我和前东德的一位供水商有过一次谈话。他说，自从引入市场经济机制，全年都有充足的水供应了，即便在旱季也有。我问他是否他们安装了新的水泵，他说没有，他们只是采用了新的收费制度并为每个用户安装了水表。从那以后，再也没有出现过缺水的情况。前东德的缺水现象是无谓的用水浪费造成的。每个户主在花园中都有一个集雨樽，樽中有一个水管。春天他们把水龙头打开，放小水流，常年流水，一直到冬季来临之前才把龙头关掉。

收费制度是把用水量控制在必要范围内的好办法，同时也是治水的资金来源。用水收费制度使供水和污水处理领域内与银行合作融资成为可能。这一点如果与持续有力的法律保障相结合，会对国际投资者更加有吸引力。在供水领域内，中国许多地方已有这样的范例。许多外国公司已经在中国投资BOT项目。

原则上，在污水处理厂建设方面也可以采BOT项目方式。这种方式的前提条件是，有持续的人口数量的信息（保证有足够的人口使用经处理的污水），因为通常情况下没有人或机构有足够的意愿垫付污水处理项目的费用。

在德国，供水和污水处理基本全部建设完毕并属于公有，不过，建设工作由商业公司组织完成并实现了赢利。这种运作模式的优点是，政府不会失去宝贵的水资源控制权，而且，政府可以根据需求变化调整污水处理的质量。

5　小结

能够对必要的项目进行国际融资是非常吸引人的事情。如果你准备通过世界银行进行融资，则必须证明你所有采取的措施已经隶属于某个项目。否则，就必须提供中央政府层面的《国家协作战略书》。为此，要计划出几年的时间。同时，世行项目必须是中央政府签订协议。

如果项目的规模不够世行项目那么大，与德国国家发展银行合作还是很有吸引力的，而且不会耗时很长。一般情况下，这种项目不会与某种供应量挂钩。但是，通常与德国国家发展银行合作需要德国进行技术和专业知识输出。与德国国家发展银行的谈判也可以由地方政府出面。

一般情况下，这些举措的目的是为了促使合作国在未来自己解决问题。为支付融资费用——这里是指水资源的融资费用，建议在饮用水和污水处理管理方面采用收费体制，以保证对某些资本投入的再融资。

在德国，市政公司很普遍，并且能够提供优质的供水管理和污水处理服务。你们能够对本国和国际的公司征收供水和污水处理费用，也能够以同样的方式运营市政公司。

WBalMo
——水资源规划模型简介与应用实例

S. O. 卡登
DHI-WASY 公司，德国

1 水资源规划——随机性问题

水的管理可以归纳到水循环中，水循环是指主要由降水和蒸发形成径流、然后进入到地表水和地下水的过程。这些自然过程掺杂着人为的成分，如地表水和地下水的利用、水资源管理措施（如水库调控和输水）等。本文将讨论在采矿区最具代表性的施普雷（Spree）河流域，同时将涉及矿井排水和矿坑蓄水的问题。

水资源管理的目标覆盖需水用户（如城市供水、电厂、工业）、维持最低径流量（因为环境生态和航海的原因）以及有效抵御洪水。

水资源管理是一个随机的问题。排水过程本身是确定性的，但缺乏控制它的水文气象过程及径流产生的时空分布等方面的理论知识，迫使我们把长系列的径流过程作为随机过程来看待。图1说明了产生径流时间序列的两个可能的方案。水资源长系列规划的模拟和分析，通常以月作为时间步长。

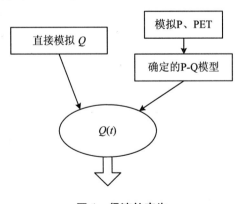

图 1 径流的产生

首先，通过采用适当的随机模拟模型方法，可以产生一定时期内不同情景下的降水及其产生径流的时间序列。第二个方面的不确定性与未来可能的气候变化有关。再一方面，虽然用户的需求很可能将取决于气候变化，但从策划者的角度来看，在时间和空间上还是确定性的。不过需水用户今后的需求由社会经济的发展确定。最后，水资源规划管理措施如水库的建设或选择，需要根据水资源的可用性和用户的未来需求量而定。图2总结了以上问题的讨论。

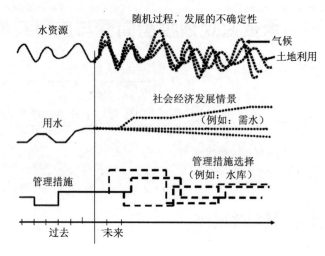

图2　水资源管理规划的各个方面

基于径流的随机性原理和水资源管理结果，水资源的随机长系列管理方法得到了很大发展，主要是在以"大需求、小水源"为特征的地区，就像在德国东部地区，尤其在施普雷河流域下游的卢萨蒂亚（Lusatia）矿区。

随机管理模型将水资源管理问题划分为三个部分（图3）：

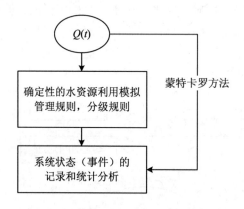

图3　随机水资源管理模型的方法

- 气象和水文过程的随机模拟；
- 水资源利用过程的确定模拟；
- 记录相关的系统状态。

如果按每月一次模拟做了足够长的时期，记录系统状态的统计分析将对水的分配情况提供足够精确的近似概率，如水库水位、特定的水平衡状态的排水以及水量供应安全限等。

2 模拟系统软件 WBalMo

基于对上述一系列模型方法的讨论，长期规划模型在最近几十年得到了发展（Schramm，1979）。这种类型的第一个广义模型（GRM）开发于20世纪80年代（Kozersky，1981）。在90年代，德国水资源规划和系统研究所（WASY）对这种模型进行了进一步改善，使其适应了新的硬件和软件技术。最近的技术是拥有 ArcView 用户界面的 WBalMo 模型系统。

基于江河流域的知识结构以及具体的天然径流和水资源处理过程，在不同条件下流域水资源系统的定量分析可由 WBalMo 系统软件进行模拟。

这种管理的模式基于蒙特卡罗方法，它是 WBalMo 软件系统的基础。它可以对流域中任何时期的月用水流程进行再现。与此相应的相关系统事件的注册，简化了模拟中完成已注册事件的统计分析。水库容量、个人用户或流动供水系统的不足等区域分布的近似概率结果适用于所选择的河流剖面。基于这一点，选定的管理策略的质量可以由研究的江河流域进行评估，这种管理策略的逐步改善可以通过对目标变量的计算来完成。

WBalMo 软件将随机性条目的大小和确定的用水过程的再现处理严格区分开。对于所选择的时期，在一个月的基础上，WBalMo 通过产生时间序列的大小来计算径流产生的时空分布。根据规则，用与排水形成的时空条件相符合的随机模拟模型来计算。

以月为时间步长的仿真过程如图4所示。

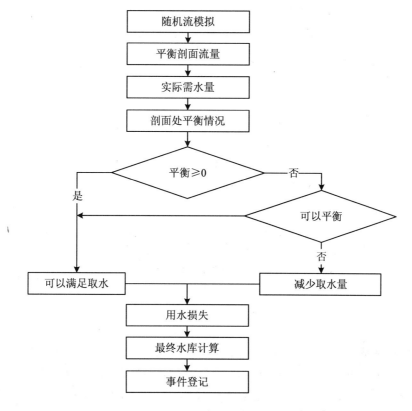

图4 WBalMo 模型的平衡过程

2.1 模型假设

以下假设构成了 WBalMo 模型系统内的水资源利用过程再现的基础：

- WBalMo 是以在江河流域中利用流动水体和平衡剖面的水文过程图示为基础的。
- 通过细分整个覆盖区，计算模拟子区域并把其赋值为以上指定的排水时间序列。然后这种排水作为单独可用的获得量分配到平衡剖面。
- 考虑用水户时要与其特定区域和数量一致，其区域和数量是从平衡剖面中分析和反馈必需的水量得到的。
- WBalMo 是基于水库（地表水库、湖泊）的位置、库容和输出要素等特质的，这些特质描述了它们当前的定向需求。

WBalMo 系统结构的例子如图 5 所示。

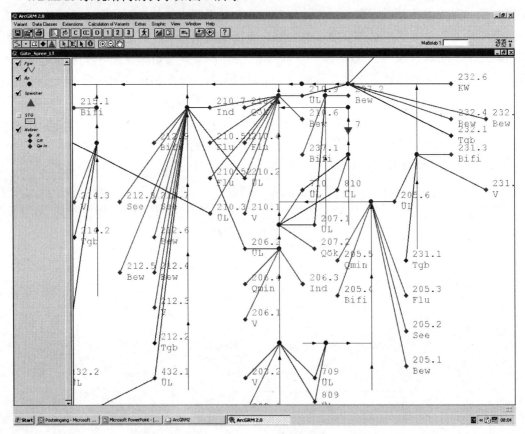

图 5　WBalMo 的模型结构

2.2 用户排序

所有的用户和输出量都被分配到一个数字序列。这使得每个用户和他们的用途可以存档到整个系统。针对不同的用户或用户组，水库的输出量是不同的，例如，在干旱期间大量用水是可以的，而其他时间用水量必须有所减少。用水过程的模拟需要遵循一个固定的运算法则——对应于不同序列号的用户对水库排水量的需求不同。

最新 WBalMo 版本的创新有：
- 程序系统中江河流域的系统略图可视化；
- 模型数据填充到数据库；
- 提供一个外部模型接口，允许在 WBalMo 水量模型基础上的特别调查（如水质、日模拟值）。

3 施普雷河流域的水资源规划

3.1 流域简介

施普雷河流域位于德国东南部，流域面积约 10.000 km²。图 6 给出一个近似的概况。图中研究的重点分别是卢萨蒂亚褐煤开采区、施普雷林山（Spreewald，一个重要的湿地地区）和柏林市。

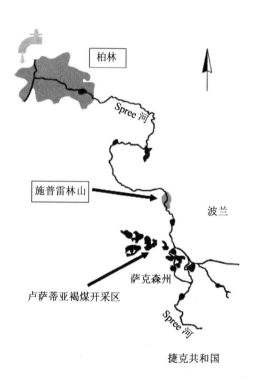

图 6 施普雷河流域概化图

3.2 褐煤开采区水资源管理中存在的问题

卢萨蒂亚褐煤矿区（Lusatian Lignite Mining District）位于柏林东南部约 80 km。由于片面的能源政治方针，原德意志民主共和国对褐煤矿的开采没有考虑生态方面的因素。为了褐煤的开采人为地降低地下水水位，受此影响的区域接近 2 100 km²。排出的废水主要注入施普雷河中。

纳污河流接收到的来水量明显增加，来水稀释了市政污水和工业废水，因此对水质产生了较小的影响。然而，由采矿引起的影响却严重地增加。由于酸性矿山废水的主要来源——黄铁矿的氧化，高浓度的铁和硫酸进入到纳污河流中（Schlaeger 等，2000）。

1990 年德国统一以后，大部分露天矿山的开采因为经济方面的原因被迫停止。因此施普雷河的排入废水量减少。此外，施普雷河及其支流的水现已用来填充露天开采的矿坑以及回灌到地下水漏斗区。由此产生的水资源短缺影响了施普雷林山湿地的最小流量和柏林等地的供水。

由于水资源短缺引起的稀释水量的缺乏也相当严重地影响了河流的水质。此外，由于矿区地下水位的升高和正在进行的矿山排水，开采矿山对水质造成的影响仍在继续。

3.3 Spree–Schwarze Elster 流域的 WBalMo 模型

为了模拟本流域高度复杂和进程相互关联的径流形成和水资源利用情况（包括 Schwarze Elster 流域），WBalMo 管理模型已经开发利用并永久地被水行政主管负责部门使用。

该模型的组成可以由以下模型参数说明：
- 170 个平衡剖面；
- 64 个模拟分区；
- 415 个用水户；
- 14 个水库和 52 个水库排水系统；
- 53 个操作系统特有原理（特殊的非标准子过程的子程序）；
- 280 个统计注册。

这个模型的应用之一是分析需水量、设计新水库的结构或是从其他区域（Odra 河流域）来水的迁移途径。采矿区的大型矿坑可能作为水库使用。图 7 说明了不同情景下的模拟结果。

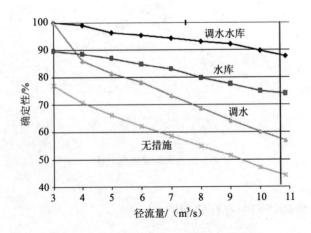

图 7 不同情景下施普雷河的排泄量

图 7 中 X 轴代表施普雷河的径流量（资料由靠近柏林的水文站获取），所需的最小流量约为 11 m^3/s，Y 轴是一定流量下的保证率。从该图可以看出：如果不采取措施，一年以

后的径流量将不能保证，每隔一年以后的径流量几乎都只有必需量的50%。不同的措施对稳定流量的影响是显而易见的，但是，即使有水库放水和跨流域调水，规定的流量也不能保证每10年都相近。

4 目前的研究进展

4.1 WBalMo 的模块化

柏林区域和施普雷林山湿地是施普雷河流域的两个重要子区域，在 WBalMo 中，它们被分别独立建模。柏林模块已经完全被嵌入 WBalMo 模型中，但是施普雷林山湿地的情况则有所不同。为了详细研究湿地系统，本文开发了独立的施普雷林山模块，这个模块虽然没有加进 Spree - Schwarze Elster 流域的 WBalMo 模型中，但它是直接耦合的。虽然不可能看到详细的计算过程，但是它展示了模型化的水系统与河流系统、用户、水库和平衡点的高度复杂性。

有重要子流域的更大的江河流域在模块化方面更重要，因为一个单一的管理模型可能变得相当大并不容易处理。此外，常用的子流域模型已经存在，它需要嵌入一个大模型中。模型之间的耦合必须是交互式的，即一个子模型中的管理决策可能影响其他模型。图 8 为德国易北河（Elbe）流域（大约 15 万 km^2）的 WBalMo 模型概化情况。

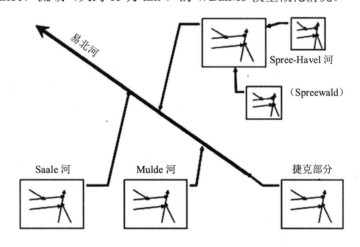

图 8 易北河的 WBalMo 模型

4.2 长系列规划和水质

正如上面讨论的，水质在施普雷河流域具有重要的控制价值。除了非点源污染主要来自于农业、点源污染来自工业和市政用水，现存的和已关闭的褐煤矿井是最重要的污染源。这些参数之间的重大转化过程和参数、沉积物和大气之间的交互作用必须要考虑。

施普雷河上游地区的第一个 WBalMo 水质模型是由施勒格尔（Schlaeger，2000）等人开发的。为了获取到所有与水质相关的引入量和描述在取样站的水资源平衡，基本系统结构必须要加强。附加的平衡剖面和新的用水户（污水处理厂、工厂、鱼塘等）必须要引入。

图 9 显示了相关的参数。

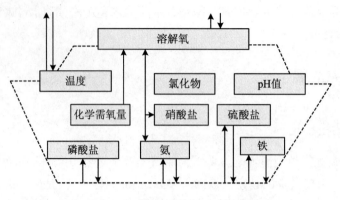

图 9　褐煤矿地区相关的水质参数

水质模型通过分别在平衡剖面、用水量、支流、非点源排水和实际的时间步长或月份中的流量 Q_i 的模拟与水平衡模型相联系。对模拟转化过程十分重要的流程时间 t，可通过定义河段的特征函数 $f(Q_i)$ 计算出。

因为在长系列平衡模型中水质的模拟比在过程模型中考虑得更粗略，因此详细的变化过程必须要简化，但是还需要十分精确地描述本质的反应。施勒格尔等人（2000）介绍了建立简化这种水质模型的方法。作为建模的例子，图 10 显示了排放的矿井废水对 COD 的模拟影响。

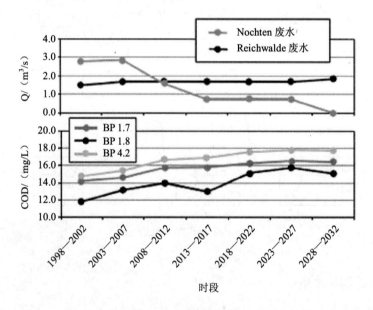

图 10　排放的矿井废水对 COD 的影响

这个模型的建立是整个施普雷河下游区域（从 Spring 区到柏林区）项目的一个试点。

参考文献

[1] Kozerski, D. (1981): *Rechenprogrammsystem GRM als verallgemeinertes Langfristbewirtschaftungsmodell.* In: Wasserwirtschaft-Wassertechnik 31, H. 11/12, S. 380-394, 415-419.

[2] Schlaeger, F., Köngeter, J., Redetzky, M., Kaden, S. (2000): *Modeling and Forecasting for Water Resources Management: The River Spree Project* Paper at Hydroinformatics 2000, Iowa, USA, July 2000.

[3] Schramm, M. (1979): *Zur Anwendung stochastischer Methoden bei der Modellierung wasserwirtschaftlicher Systeme.* Dresden (Habilitationsschrift, TU Dresden).

Urban Water Management
– A Global Challenge

W. F. Geiger
University of Duisburg-Essen, Germany

1 Misperception of unlimited availability of water

Water is one of the rudimentary needs for daily life, agricultural and industrial production. Renewable water is limited in many regions and deep groundwater resources are considered not renewable in generations' time. Still, water is much more abused compared to other natural resources. Man uses water as if its sources were unlimited. Philosophy, science and technology all contributed to this misperception.

Looking back to the history of human civilization, cities always preferably settled at rivers, allowing direct use, trade, bringing wealth and well-being to people. Ancient cities like Miletus being not able to access deep groundwater practiced sustainable management of the renewable resources integrating art and architecture in water supply and wastewater disposal structures and fulfilling social needs. Thales of Miletus (624 to 545 BC) established theories on water cycle, which later were discussed by Platon (427 to 347 BC) and were finally explained in a scientific way by Palissy (1510 to 1590) almost two thousand years later. The theory of a continuous water cycle suggested that water constantly becomes available by rain. There was always water flowing in the streams from often unknown upstream regions and natural springs brought groundwater continuously to surface for human use.

Due to the availability of electric driven pumps after the industrial revolution in the 19th century sustainable water management was neglected, after even in dry seasons water could be augmented. By water always flowing out of the tap where public water supply systems existed, public awareness for the natural variations of water availability got lost. Even in rural areas where public water supply systems often not exist, deeper and deeper groundwater resources were accessed. Water supply seemed guaranteed and little efforts were made to focus on water harvesting in rainy seasons. In contrary during flood periods, water was carried away as much as possible and eventually discharged to oceans.

The aforementioned philosophy of a continuous water cycle, the apparently ever repeating rainfall and seldom dry falling springs and rivers, the easy accessibility of water with the help of modern techniques, left people an impression that water resources are unlimited, a perception which is reflected by the water use behaviors, especially in regions where water is scarce. For example, leaking and running taps are not rare instances. Water becomes a commodity that everyone uses and disposes without thinking of its origin and its destination.

Today's industrialization and urbanization basically is on the account of future generations. Accepting deterioration of the environment for economic growth and then trying to repair environmental damages with the economic gain repeats one of the biggest mistakes in history. Thus there is no justification for developing countries to damage environment for economic growth only because industrialized countries made same mistake earlier. Even worse, in developing countries urban growth occurs before economics allow for the necessary infrastructures. Only a few cities have resources and skilled personnel to supply the fast growing population with water and sanitary installations. The unhealthy environment in slums leads to social depression and crime. Even more the problem is political as water problems usually are solved only after streets, energy and communication lines are installed, while water should be taken care of first. Urban water problems will increase especially because in a few years for the first time in man's history the majority of earth population will live in cities (60% in 2025). Urban agglomerations in South America, Africa and South East Asia will double between 2000 and 2025.

2   Challenges ahead

2.1   Water scarcity

Centers of civilization have always been centers of water consumption. Where local resources were limited, water is transported from near and far regions to the city. The larger cities are, the more they dominate water resources of the whole regions often ignoring the need of local populations. It is hard to understand that for instance in Beijing (in the last century, not in ancient times) clean water was abused for air conditioning, car washing and street cleaning and that large irrigation facilities were installed to sprinkle lawns and at the same time rainwater was disposed out of the city in huge channels at a time where groundwater dropped at a rate of two meters per year. It cannot be justified that at the Mediterranean Sea in Europe a tourist consumes in summer 1 000 L of water per day where water principally is scarce. Abuse of water occurs everywhere in the world disregarding water scarcity, which is accelerated by the rapid expansion of cities under the global urbanization process.

By changing surfaces from green land to houses, paved streets, plazas and yard ssurface runoff is accelerated and groundwater recharge is inhibited. To ensure safety of the citizen, storm water is expected to be discharged immediately into sewers and urban rivers, which lastly flow out of the cities. Consequently it is unavoidable that water resources deplete to an extent where the natural water cycle irreversibly is disturbed. At the beginning the disastrous consequences of urban growth and of exaggerating water consumption were not realized, as ever new techniques allowed accessing new sources such as deep fossil groundwater. However, fast sinking groundwater tables have been reported in many cities due to overexploitation of groundwater resources.

2.2 Water pollution

Pollutants inconsiderately are discharged to soil and water bodies making them unusable. The pollution problem often is transferred to other regions and to the sea by groundwater movement and rivers. Little recognized are the chemical time bombs which are stored in soil and sediments and passed on to future generations. Land use and climatological changes may remobilize toxics and later endanger lives.

2.3 Water price

Water often for political reasons priced lower than its costs enhances sloppy water use. The user does not realize the true value of water. In a small community in India for instance, where water resources are extremely scarce water, is supplied from far distances at high costs, but made available at no cost. Asking people why the public taps never were shut off during times when nobody collected water the answer was: water does not cost anything.

2.4 Limitation of current water management methodology

Use of water was not foresighted in the past. Problems resulting from overuse often were not recognized in time, not all problems recognized produced enough pressure to be dealt with by politics, not all problems politically treated led to decisions, and not all decisions led to action. End of pipe measures often were taken by case to case decisions, close to the damage, but far away from the source of problem. Actions followed the enemy approach, i. e. get rid of storm water and used waters as quick as possible and dispose it out of the city. Actions taken only solved particular problems and were not integrated into holistic plans. As a result it could not be avoided that groundwater tables dropped, flooding increased and natural habitat disappeared.

2.5 Emerging problems

In many regions a new phenomenon occurs, namely that mega cities grow together as it happened already in the Pearl River Delta between Hong Kong and Guangzhou, in the Jakarta-Surabaya-corridor and Japan's Tokaido corridor between Tokyo, Nagaya and Osaka. The aforementioned water related challenges get even worse in the urban agglomeration areas. Here already today it is obvious that water problems cannot be solved with conventional techniques for water supply and wastewater disposal. The new problems of mega cities together with the pressing inheritance of the 20th century, namely that one billion people are not sufficiently supplied with drinking water, two billion people have no adequate sanitary installations and four billion people pollute water which cannot be treated adequate, cause a tremendous burden. The problem is not reduced neither by accepting that this situation is man-made nor ignoring the consequences of inadequate water use.

Climate change creates significant impacts on freshwater resources, with wide-ranging consequences for human societies and ecosystems. Forecasted increasing precipitation intensity and variability increase the risks of flooding and drought in many areas. Many forms of water pollution are projected to be worsened due to the changes in extremes, including floods and droughts. The water management practices under current situation may not be forceful enough to cope with the above mentioned impacts of climate change (Bates et al., 2008).

2.6 Failure of attempts to incorporate social and ecological aspect into water management

In addition to the function of supplying drinking and production need, urban waters have also cultural, esthetic, social as well as ecological function. Already in 1970 Professor McPherson, one of the early urban hydrologists demanded that water management must include social and economic aspects. By current piece meal technical actions, one never can reach technically and economically sustainable solutions. Obviously, social peace and ecological balance is in danger then. Further, the quick change of employment and social structures can ruin the achievement and lead to unemployment with its social consequences and disability to deal with hazardous wastes of past industrialization.

3 Call for local solutions

3.1 Need for comprehensive water management

Increasingly political decision makers are aware that economic development and the condition of the environment and the availability of water cannot be separated. Ignoring the carrying capacity of water limits economic growth and destroys the results achieved. In this sense poverty at the same time is the main cause and the main consequence of urban water problems.

Already due to globalization cities compete world-wide for new industries and economic growth. The economic development directly is linked to the availability of water and the costs for its supply. Sustainable development requires a change in mind. Water consumption must depend on the local availability of water. Water should not be supplied on the cost of environment. One should distinguish between the elementary basic need and additional needs. A splendid historical example is the water distribution in the Roman city of Nimes, in which the pipelines for public water supply, commercial water supply, swimming pools and private houses were stepped at a distribution chamber in a way that public water always was provided. Other users were served in above sequence only when enough water was available. Such in times of water scarceness only the public basic needs were covered.

3.2 Need for decentralized responsibility and action

Balanced water management requires that the natural conditions, in specific renewable water resources are connected with social and economic needs and desires. Development of water concepts cannot be based on predetermined solutions. Solving the water problem requires a typical solution for each city, each part of the city and even for each neighborhood. What functions in one city may be fully inadequate for another city. This insight forces the decentralization of responsibility and action. Normally the costs for large-scale technical water supply and disposal systems are higher than the costs for small autarkic systems. Operation of small units also is less vulnerable. Smaller units involve small and middle size undertakings for construction and operation and thus strengthen social-economic structures. Small autarkic systems also more easily remain functioning, as the responsible people and users identify themselves with the system. Water neighborhoods can carry more responsibility for preventive measures. The condition, however, is a local feeling for belonging together, which is difficult to achieve when the population differs in its cultural, religious, political or social status.

3.3 Adjustment of administrative structure

It cannot be overlooked that conventional structures of water authorities oppose the introduction of site specific decentralized water systems. Central water authorities usually are subdivided into different responsibilities for irrigation, drinking water supply, urban drainage and wastewater treatment. Central authorities usually concentrate on meeting water supply and disposal requirements and are unable to deal with the even more important sensitive use of water. Differences in uses are not differentiated, which leads to high costs for supplying water of one unique quality. Administration is mainly focusing on more or less central systems. However, in both supply and drainage, it has shown that half central system, even decentralized system, also in urban areas has significant economical and operational benefits. Also, administration structure shall be changed to cope with local situation.

3.4 Use of local resources

Water in cities also calls for dual systems. It is important to make use of local resources such as rainwater, to reuse greywater and use rainwater as much as possible for irrigation and recharge of aesthetic ponds in cities. All of this also helps to improve local ecology. Water supply shall try to make use of local sources first, which makes the neighborhoods more independent from external water supply. Water neighborhoods themselves can determine their social identity and adapt water use to their tradition and cultural needs. Citizens usually are more creative than their governments by taking initiative in time to prevent upcoming problems. Therefore, public participation is strongly encouraged for local water related decision making to discovery and utilizes the local wisdom.

3.5 Local systems are the future for megacities

Local systems have a good chance to succeed as even in the largest mega cities as people still are living in small neighborhoods. They come to their home, use local shopping possibilities and cultural activities. Therefore already today the citizens independent of the size of their city normally only live a limited local life. The city in the future must be a city of neighborhoods, in which on a small scale living style and development is determined. Central water authorities will regulate and control the water business carried out by commercial regional outfits. This concept of course calls for discussion of the obvious discrepancies between "economics of scale" and "small is beautiful".

The question is how long such a change will take. One may remember that Hesiod about 800 BC pointed out the elementary connection between water pollution and health of man. At that time it needed 300 years until sanitary measures were installed in Greek cities, although they existed in previous cultures such as the cultures of Mesopotamia and along the Indus River. It will be a learning process which begins in schools and continues throughout life. The renewal of thinking about water must come from inside, from the user itself.

4 From local solutions to global thinking

The local solution to the problems does not exclude global thinking, especially as local water problems around the world have a lot in common. Hydrologic and technical principals can be used everywhere. Guidelines for the local function of water units can be established in general terms as well as recommendations how cost intensive but ineffective measures can be changed to cost saving and effective solutions. Technologies which help to save water can be used everywhere. Worldwide the global trend for increasing water demands, abuse of water and mismanagement can be counteracted. Globally it also can be regulated that all users must be involved, that river basins must be considered as hydrographic and economic units, that river basin wide water management plans must be established leaving option for local action and that water prices paid by the users must reflect real costs.

Ecosystems do not stop at national boundaries. This enforces an international coordination of water resources management. In a global network regulations have to be agreed upon which, however, leaves sufficient freedom for regional and local action. It would be wrong to understand globalization as standardization. On a global level the basic principles must be recognized and regulated, on a local level suitable solutions must be found.

Global thinking should lead to solidarity to deal with water, whereby the decisive responsibility for coordinated action lies with the industrialized countries. These countries must begin to introduce the new structure as they usually are in a better economic condition then developing countries. They had developed in the past on the account of nature and environment and therefore carry the responsibility to show better ways for the future. Declarations of intentions and ideas are as little sufficient to protect water resources as recommendations for developing countries are to protect their resources. The environmental crises of industrialized countries in the second half of the last century by far are not mastered. Therefore also industrialized countries can benefit from the new manners to deal with water. The global economic competition between cities must change to a global competition of good ecological conditions, which at last again will lead to sustainable and competitive economic structures.

5 Vision

We hope that the honored reader in the year 2030 will notice when he looks back, that the vision became reality: what started in the last quarter of the 20th century with pilot projects which proposed local and independent water concepts securing basic water needs and managing local water resources according to natural patterns protecting ecosystems and increasing the urban environment were accepted as regular task in the first decade of the 21st century. The renewal was accompanied by a change in mind, which did not consider water as a matter to use and to throw away anymore but as a valuable good which was dealt with carefully such as if it was borrowed from future generations. The true value of water was recognized by everyone. The globalization of markets and media spread the thought very fast. Increasing water problems, which were left by the 20th century, were solved.

In the second and third decade of the 21st century in all cities water neighborhoods were established which formed their surroundings in common responsibility. Each household focused on saving and protecting water. Urban and regional private institutions were linked into river basin wide water management and responsibly supplied the local water neighborhoods with fresh water and treated surplus wastewater for further reuse. Slowly a network of water trading between water neighborhoods, cities and surrounding agriculture established. The responsibility of water authorities for managing the natural resource water was fully separated from the water supply and disposal industry which only was economically based.

In the declining third decade of the 21st century trade of water became a matter of course. Water was priced according to its true value. State water authorities only had regulative and control functions and coordinated their regulations internationally and globally. The global network of independent water suppliers and local water neighborhoods increasingly functioned. The water neighborhoods directly helped each other to solve local problems. Citizens took full responsibility for their water and surroundings. Urban habitat improved everywhere and cultural water-related traditions revived. Nature came back even into mega cities and provided shelter for emotional and physical existence. The dual water system proved to be efficient and led to sustainable development which not only satisfies the need for the present generation but also leaves all possibilities open for future generations.

This condition was achieved by the insight and foresight which mankind gained at the turn of the century. The whole world in the second half of 21st century is an urban network which is able to react locally fast and has a chance to survive. Already forgotten are the forefathers which paved the way for this development by conducting pilot projects in the last quarter of the 20th century.

This paper was primarily based on two earlier publications of Geiger (1993 and 2001).

References

[1] Bates, B.C., Z.W. Kundzewicz, S. Wu and J.P. Palutikof, Eds., (2008): Climate Change and Water. Technical Paper of the Intergovernmental Panel on Climate Change, IPCC Secretariat, Geneva, p. 210.

[2] Geiger,W. F. (1993): Urban Drainage - a Technical, Environmental or Social Problem?[C]//Proceedings of the International Conference of Aquapolises, Shanghai, November 17-19, 1993.

[3] Geiger,W. F. (2001): Global denken und lokal handeln[C]//WaterScapes – Planen, Bauen und Gestaltenmit Wasser.

Sustainable Water Management
– From objective to implementation

W. F. Geiger
University of Duisburg-Essen, Germany

1 Introduction

Many countries, also China face the dilemma of rapid industrialisation, fast growing population in cities and increasing water demands for urban living and lifestyle as well as for irrigation of green spaces. Worldwide water demand has increased six fold between 1900 and 1995 though population only has doubled. For this water resources are stressed, rapidly exploited, ignoring degradation of soil and groundwater tables. This creates new problems, hardly envisioned even a few decades ago. More and more it is recognised that old conducts of water resources management must be significantly modified in order to cope with continuing growth in the 21st century.

In specific, water resources management in the past suffered from the difference between the long times, often. up to ten years, which usually are required to assess current conditions by monitoring, to estimate the water quantity and quality situations in a river catchment and to identify development and environmental protection strategies, and the short times, often less than one year, available for decision making to proceed with local action, necessary due to quick changes in urban and industrial development. Further, regulations and guidelines lack behind the availability of new technologies for water management. In consequence quick decisions cannot be checked if they fit into long-term water management intentions. This may result in low efficiencies and high costs. So far there is no unique procedure that allows linking long-term planning and short-term implementation of local actions. In specific quick and sufficient assessment of problem areas and a manageable overview on effectiveness and costs of technical and non-technical abatement measures are needed. Vibrant techniques are lacking to link water management to sustainability.

Water management in future also has to pay special attention to the merging of urban agglomerations into mega cities, as it already happens today in the Pearl River Delta between Hong Kong, Shenzhen and Guangzhou, in the Jakarta-Surabaya corridor and Japan's Tokaido corridor between Tokyo, Nagaya and Osaka. It is expected that 60% of the world's population will live in cities by 2025 and 90% of the urban population in mega cities.

2 Principles of sustainability and water management objectives

Sustainable water management aims to protect water resources and ensures that also future generations have sufficient water. Population growth and desired economic development especially in regions with unfavourable hydrologic and geographic conditions make it difficult to find solutions that not deteriorate water resources in a way that they burden future generations and still are acceptable in a socio-economic sense. Regarding the need for integrated approaches international organisations i.e. UNESCO through their international hydrologic programmes (IHP) already in the 70ies of the last century tried to bring together meteorologists, hydrologists, biologists, social scientists, economists, agrarians, city planners and others to improve the understanding of the many facets of water management and to disseminate the knowledge to developing countries (i.e. UNESCO, 1987; UN, 1994).

2.1 Holistic Approach – conditio sine qua non

Sustainable water management recognizes the system complexity and interconnectivity of its elements. Holistic in its approach, it equally involves local and regional authorities, employers and scientists, environmentalists and decision makers, politicians of all parties, governing and in opposition, and especially the people affected. Sustainable water management ensures that no substances are accumulated or energy is lost by recovery and reuse techniques. Sustainable development is "development that meets the needs of the present generation without compromising the ability of future generations to meet their own needs" (World Commission on Environment and Development, 1987 – "Brundtland report"). It also ensures that future generations will have the means to achieve a quality of life equal or better than we have today. In other words sustainability avoids future regret for decisions taken today. Further sustainability implies that consumption and production in one community should not undermine the ecological, social and economic basis for other communities to maintain or improve their quality of life. Last not least it conserves biodiversity and takes precautionary risk aversion strategies.

Water management is cross-sectorial by its nature. The division of the society and its institutions into various sectors becomes an obstacle in achieving the goals of integrated water management. The creation of inter-sectorial links, supporting cross-sectorial co-operation and integrated multi-disciplinary actions, represents the greatest challenge in the implementation of sustainable water management. This is made difficult by the current sectorial educational systems which are not designed to impart the broad integrated knowledge necessary to promote a multi-disciplinary, holistic approach to resource management problems. In consequence, sustainable and integrated water management needs novel administrative approaches and holistic education.

The most important environmental/ecological goal is the protection and enhancement of ecological integrity. For this, space and time are important. The species present at any time in an area depend on what has happened in earlier times. Thus the management strategy cannot focus on present conditions, but must consider past history and the current characteristics of organisms present.

In defining socio-economic objectives, one must consider cultural, geographical, economic and political conditions. Socio-economic objectives are concerned with supply needs, flood protection and preserving natural environmental conditions. Without question, sustaining and protecting human life is the most important socio-economic objective of water management. But also maintaining diversity of the ecosystem in at least parts of agricultural and urban areas is an important socio-economic goal. Many developed areas have obliterated the natural system and now must incur great costs to take remedial action. To quote an old saying: "an ounce of prevention is better than a pound of cure".

The ability to quantify, in monetary terms, the socio-economic dimension of an urban water management project does not determine its importance. Analysis should differentiate between price and value in situations where price is a current, transient factor, and not an accurate measure of ultimate value. The conventionally used monetary costs should be supplemented by environmental costs, which consider positive environmental effects as well as environmental damage caused by implementation of environmental measures. This is needed for attaining sustainability of water resource management.

2.2 Water Management Objectives - status in quo

Safe, reliable and equitable water supply is one of the most important objectives of urban water management. In urban areas, concentrations of population, properties and economic business activities are particularly high. Consequently, the objectives of protection against harmful floods, and particularly against the loss of human life, receive the highest priority. Sanitation and protection of surface water quality are very important objectives as well, not only from the point of view of environmental protection, but also from a public health point of view. Obviously water management has to meet different and often contra dictionary goals to simultaneously fulfil its economic, social and environmental goals.

The different components which make water management sustainable do not receive the same priorities at all times. At different stages of societal development, different objectives receive more importance. In the pre-industrial society, emphasis was placed on drinking water supply, transportation and water supply for irrigation. In the industrial society, generation of hydropower and waste disposal and transport are prioritized. Finally, in the post-industrial society, high emphasis is placed on aesthetics and ecology. Water-based recreation within urban areas is widely practised in industrial and post-industrial societies. Figure 1 sketches the interaction of industrialization reflecting priorities set for water management and the environmental condition. While the existence of changing priorities must be recognized, at the same time lower priority uses cannot be neglected over a long run of time, because of interdependency of various uses.

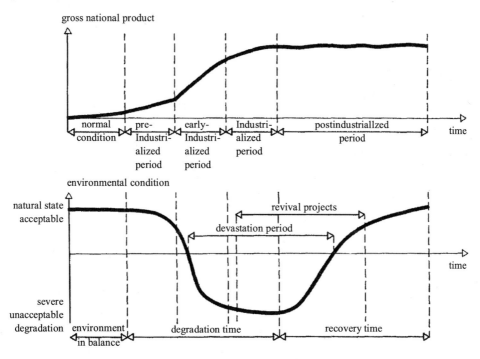

Figure 1   Interaction of industrialisation, economics and environmental condition (UNESCO, 1995)

2.3   Importance of spatial extent and time consideration

Sustainable development not only is multidisciplinary but also multidimensional as it has to cover short to long periods of planning and it has to address small-scale problems as well as regional problems. It even includes a global view.

The assessment of the water situation requires at first the definition of the catchment boundaries. This is not an easy task, as the definition of the boundary also depends on the problems to be solved. To reduce ammonia toxicity or oxygen depletion downstream of an urban area for instance it may be sufficient to limit emitted pollutant concentrations and thus to assess the situation within the urban area only. If the problem is nitrification or sedimentation in far distant water bodies agricultural inputs as well as urban discharges have to be assessed within the whole contributing area. This is identified in Figure 2. Further planning of abatement measures can be restricted in the first case to the emission sources within an urban area while in the second case all possible interactions with different land uses must be incorporated and the most cost-effective measures must be taken.

While for the evaluation of the effect of planned measures within urban areas detailed models reflecting the rainfall-runoff and pollutant processes are applied, in the second case mainly empirical methods are necessary which have a lower requirement on data compared to physically based methods. As Figure 3 shows the spatial extent of the catchment, the time span to be considered and the technique applied to establish water and matter balances are strongly linked. Usually data availability is a limiting factor for water management. Furthermore, if hydrologic catchments are shared by different countries it is very important to apply the same procedures for assessment and to consider the catchment as a whole.

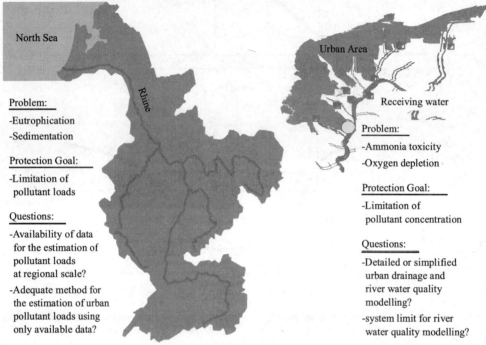

Figure 2   Protection goals of different surface waters and related questions (Nafo and Geiger, 2004a)

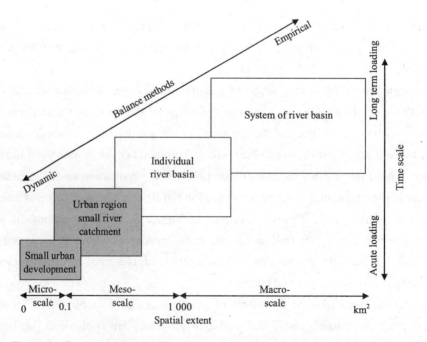

Figure 3   System limits and time horizons for water management (Nafo, 2004)

3 Flexible decision support system for water management

For water management in Europe the European Water Management Directive provides a framework comprising environmental, social and economic goals. Such framework cannot prescribe firm procedures how to meet the goals. Therefore a flexible procedure is suggested that allows for adaption to regional needs and for quick identification of problem sources as well as of cost effective measures, but also for intensive problem assessment by monitoring and modelling of management strategies.

3.1 The European Water Management Directive

Water always has been one of the major concerns of the European legislative. So far 25 regulations exist that address surface waters, fishing waters, bathing waters or drinking water quality in specific. Also limiting values for emissions from different water uses have been specified. The new European Water Management Directive (EU, 2000) provides general objectives for all waters, surface water and groundwater. As general objective a good status of water quantity and quality is demanded specifying a deadline, when the good status has to be reached. Further it is defined that water has to be managed on a river basin wide scale and combined approaches for limiting emission and meeting immission standards have to be followed.

The environmental goal is that in future further deterioration of the ecologic condition in surface waters has to be avoided and even more until 2010 a good status of all surface waters and groundwater has to be achieved. Socio-economically water prices for all uses are demanded that reflect real costs. For water management decisions utmost participation of the population in water management decisions is requested. As in Europe natural conditions are diverse, a generalised standard for biological quality of water was not given. The goal therefore only was described by "good ecological status". In consequence the biocenosis in water bodies should deviate as little as possible from natural conditions. Therefore in a first step the biological quality in a water body has to be assessed and compared with the biological quality of a natural reference condition. Such comparison will yield physic-chemical and hydro-morphological indicators that would be necessary to achieve a good ecological status. In this context the effects of urbanised areas, agriculture and other human activities on water bodies have to be assessed identifying pollution by point and diffuse sources, water extraction and wastewater discharges.

For each river basin programmes have to be installed to constantly control the status of water bodies. For surface waters such programmes include ecological and chemical measurements, for groundwater the chemical condition and quantitative development has to be monitored. Further in each river catchment an economic analysis of water uses has to be conducted, including figures on water extraction and distribution, collection and disposal of wastewater, corresponding prices and costs, allocation of costs to different economic sectors and long-term estimates for water availability and demands.

When implementing the European Water Management Directive certain exceptions probably will gain importance and political attention in future. That is that in heavily urbanised and industrialised areas as well as in areas with intensive agriculture and animal husbandry between normal changes and severe changes (heavily modified) is distinguished. In consequence the deadline for achieving a good status (2010) may be prolonged if the good status only can be met with proportionally high costs as related to the ecological, social and economic gain. The goal of a good status even may be reduced if it impossibly can be met or the costs would be proportionally high, as it may be the case for instance in the heavily urbanised and industrialised Emscher region, a small river catchment of 870 km^2 in the North-West of Germany.

3.2 Steps to achieve sustainable water management

The difficulties in specifying firm goals and procedures and the regional differences in water resources and geographic conditions require a flexible procedure and possibilities to also adapt objectives if the costs necessary to repair a certain condition cannot find social acceptance. Figure 4 structures water management into consecutive steps.

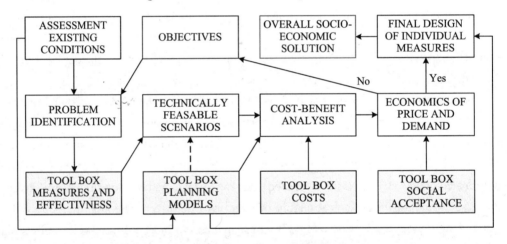

Figure 4   Consecutive steps in water management

Water management starts with assessing the hydrologic, geological, environmental, socio-economic and other relevant conditions and comparing these conditions with objectives. The objectives may be general, i. e. minimization of flood damage or maximization of amenity value or they may be quantified by water laws and regulations, i. e. specifying water quality levels to be achieved or limiting pollution concentrations discharged. While in the past for assessment extensive monitoring and surveying programmes were conducted today in some cases water quality problems thoroughly can be assessed by using satellite data. Such an assessment only may takes few weeks.

Combining for instance satellite pollution assessment with deterioration catalogues, which summarize observations in the past on pollutant sources from agriculture, animal husbandry, urban runoff etc. the problems in a river catchment quickly can be identified in a first step. The larger the problems are, the easier such identification is. In addition monitoring programmes should be established in order to allow later on to calibrate mathematical models used in the further planning process.

The next phase in water management planning is to arrive at cost effective combinations of measures to alleviate certain problems. Today there is a pronounced tendency to use all kinds of models both for showing quantitative and qualitative effects. However, mostly it is overlooked, that all models have limitations and only can be as good as the input data available. Often the quality of planning data not justifies the efforts to do modelling. In rare cases of course where interactions of measures are so complex, that they cannot anymore be identified by linking the measure catalogues only modelling of processing in interactions with the goal of finding the least cost combination of measures is justified. However, this is time consuming and expensive. Today there are a large number of models on the market which arises for the potential user the question, which is the best for his purpose.

Even more modelling results frequently camouflage the truth. Therefore in a first instance it is suggested to establish toolboxes of different measures which include their effectiveness as well as implementation and operational costs. Such catalogues include both technical and non-technical measures. An example for such a toolbox matrix is given in Table 1, while Table 2 exemplifies some selective measures that can be taken for urban water supply and drainage. From these toolboxes different scenarios to abate a problem can be derived. Finally different scenarios of measure combinations should be compared for their environmental and resource costs.

Defining resource and environmental costs appropriately not only takes economic experience but again time. If the problems identified are significant usually it is better to immediately start with implementation of the most effective measures and conducting the necessary economic investigations parallel. However, if type of costs of measures do not find social acceptance politics have to decide whether they want to change their political priorities or lower the objectives.

At present there are no generally and applicable decisions support systems for water management. For individual river catchments, such as the Rhine, Danube or the Elbe river large research efforts were undertaken which despite of all efforts showed that ecological and socio-economic components are not sufficiently linked into modelling natural processes.

Last not least it should be mentioned that water management planning by itself will not solve problems but only helps to identify the most socio-economic comprise to achieve a good status in surface and groundwater. More important it is to implement measures and to enforce laws i. e. controlling industrial or other emissions.

Table 1 Selected example for toolbox matrix for urban areas

Water management measure	Purpose/objective				
	Conserving (saving) water	Maintaining water balance	Flood control	Control of stream erosion	Stormwater pollution control
Water-efficient appliances	P, H, A	P, H, A			
Roofwater tanks	P, M, C	P, M, C			
Greywater reuse	P, H, E	P, H, E			
Wastewater reuse	P, H, E	P, H, E			
Green rooftops			P, H, E	P, H, E	
Porous paving			PS, M, A	PS, M, A	
Aerial infiltration			PS, H, C	PS, H, C	
Infiltration swales			PS, H, C	PS, H, C	
Percolation trenches			PS, M, E	PS, M, E	
Aquifer storage & recovery systems	PS, H, E	PS, H, E			
Groundwater recharge facilities		SD, M, E	SD, L, E	SD, L, E	
Erosion & sediment control				D, H, A	D, H, A
Filter strips					SD, H, C
Responsive street layout					SD, H, C
Stormwater tanks					SD, H, E
Ponds and wetlands					SD, H, A
Soil filter basins					D, H, E
Landscaping					

location
P = Lot level
S = Subdivision level
D = Division level

efficiency
L = low
M = middle
H = high

costs
E = expensive
A = average
C = low

Table 2 Selected techniques for best management practice

domestic water saving techniques	
	water saving faucets are available maintain a defined flow rate independent from pressure variations can be directly inserted into a faucet or showers
	urinals and toilets without water flushing are available due to the smooth surface coating the urine is drained residue-free into the siphon the surface of the urinal bowl is covered with biodegradable disinfections the surface coating has to be renewed up to four times per year depending on the frequency of use

- greywater recycling is successful in reducing water consumption
- systems with and without biological treatment are available
- high level of supply safety
- low noise functioning in service

domestic rainwater harvesting

- filtering is necessary prior to storage of rainwater
- self cleaning filters are available
- applying new techniques about 90% of the runoff reaches storage
- research concentrates on removal of dissolved matters

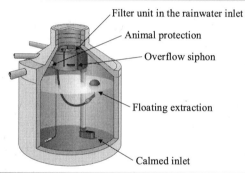

- treatment by sedimentation, floating extraction and overflow siphon
- filter unit can be integrated in the storage tank
- control only once per year
- cleaning only after 5 to 10 years necessary

decentralized stormwater infiltration

- pervious surfaces avoid runoff at source
- excellent treatment of stormwater at infiltration by vegetation
- simple and accessible maintenance
- clogging can be avoided by using approved products

- surface or subsurface storage/ infiltration facilities increase groundwater recharge
- entering water must be free from suspended or settled solids
- underground storage minimizes area requirement

centralized drainage components	
	central systems require long and large drains therefore half-central and ecologic stormwater treatment options are recommended removal of solved and dissolved matter clogging can be prevented by planting crane brake vegetation
	dependent on flood protection level centralized measures become necessary large space is necessary for flood protection multi-use of the facility space is possible, but dangerous tremendous volume are necessary (i.e. 1 000 m³/hm²)
	immense sizes of treatment facilities are needed in large cities, if treatment is not decentralized cost-effective level at decentralisation must be found operation of large treatment plants is costly wrong operation of the plant may cause water quality impairment

3.3 Flexible implementation and control procedure for urban water management

Especially in South East Asia urban areas grow very fast. Therefore it is almost impossible to make precise predictions of urban development which allow for coordinated long-term planning of pollution control facilities both for wastewater and storm water treatment. In consequence it is recommended to design elementary abatement measures for a short-time forecast, i. e. 5 years and allow for later improvements. The advantage is that once the facilities are found to be not sufficient then technologies at the time can be added.

For urban applications also present conditions have to be assessed in more detail as shown in Figure 5. For sizing facilities estimated loadings for the planning horizon have to be established. Facilities have to be designed according to existing guidelines.

As a next step either with manual methods or mathematical models emissions for the planning horizon have to estimated and compared with immission criteria if these exist. In case emission criteria are not obeyed the design has to be adapted accordingly. In case immission criteria do not exist the measures should be constructed and their effect should be controlled by monitoring. If the effects are not as good as expected at first operational corrections should be made. If these are not sufficient the facilities have to be upgraded. The whole planning, implementation and control process is shown in Figure 5.

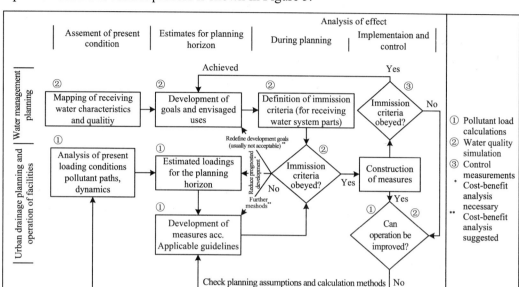

Figure 5   Interaction of drainage concepts and pollution control in receiving waters

4   Assessment of nitrogen loads from point and diffuse sources in a small river basin

According to European Framework Directive (EU, 2000) river management plans are requested at scale of river basin unit, i.e. for large river basins. But, the impacts as well as management measures have to be investigated at meso-scale level, i.e. in small river basins. Regarding nitrogen inputs, point and diffuse sources can be seen as important human pressures on surface waters. Measures against these pollution loads to achieve the environmental goals aim at avoiding long-term impacts on distant aquatic systems, like the eutrophication of seas and coastal areas, and at avoiding acute ammonia toxicity in receiving waters. The following example shows an analysis of the contribution of point and diffuse sources to nitrogen emissions into a small river basin (Dhünn river basin, Germany), and to assess their significance regarding the impacts on distant aquatic systems and on the receiving waters. For this purpose, mass balances were carried out with different methods to estimate the nitrogen loads from point and diffuse sources, the dynamics of the diffuse nitrogen emissions and also the impacts of these emissions on the upstream and downstream receiving waters.

The catchment of the investigated Dhünn river basin has an area of approx. 82 km^2. About 16% of the basin is urban area. Arable lands represent 39% of the catchment area. The main point sources of pollution are the drainage areas of two cities, one situated in the upper basin area (3.3 km^2), and the other (3.2 km^2) located downstream. The drainage areas are drained mainly by combined sewers.

4.1 Planning tools

The lumped model MONERIS (Modelling Nutrient Emissions in River Systems) (Behrendt et al., 2000) is used to estimate the nitrogen loads from point and diffuse sources. Whereas the assessment of the dynamics of nitrogen emissions from diffuse sources requires a time discrete model. The SWAT model (Soil and Water Assessment Tool) (Arnold et al., 1990; Neitsch et al., 2002) is therefore used to assess the dynamics of nitrogen emissions from diffuse sources. Acute ammonia toxicity in the receiving waters is assessed according the simple method of BWK-M3 (2001), which is based on mixing approach and the approach of Emerson et al. (1975) to calculate ammonia concentration in receiving waters. A threshold value of 0.1 mg NH_3-N/L is set as criteria for acute ammonia toxicity in the receiving waters that must not be exceeded as a consequence of sewer overflows.

4.2 Nitrogen loads from point and diffuse sources

The estimated nitrogen emissions into the Dhünn river basin and the contribution of different pathways and different sub-catchments are presented in Figure 6. Assuming a deposition value of 25 kg/hm^2·a the total nitrogen emissions into the Dhünn river basin are estimated to 270 t/a in the period 1996—1999, without nitrogen loads from the lake. The estimated loads including the nitrogen contribution from lake and the nitrogen loss in the Dhünn river system differ by only 2% from the observed loads.

Diffuses sources (arable lands) are identified as the main source of nitrogen emission into the river basin. Emissions from diffuse sources represent 79% of the total emissions into the river basin. The contribution of point sources to the total nitrogen emission is about 21%, whereas 18% of the nitrogen emission is generated by waste water treatment plants and 3% by sewer overflows. However, in some sub-catchments the contribution of point sources to the nitrogen emission reaches 35%, including a contribution of sewer overflows of 8%.

According to these results, one may lead to the assumption that point sources, especially sewer overflows, are negligible for the protection of distant aquatic systems. But, the interactions between sewer and WWTP as well as the background nitrogen contribution and land use distribution should be considered for the assessment. Sewer and WWTP should be regarded as a functional unit. The background contribution to total nitrogen emissions is about 43 t/a. Thus, arable lands generate nitrogen emissions of 171 t/a, which represent 54 kg/hm^2·a. In comparison to that, emissions of about 56 t/a (42 kg/hm^2·a) are generated by urban areas. These results mean that urban areas, that represent 16% of the investigated river basin, cause 21% of the total nitrogen emissions. Whereas arable lands, that represent 39% of the investigated river basin, cause 63% of the total emissions. It can be concluded that the contribution of point sources to the nitrogen emissions is not negligible in the present study. Thus, measures to protect distant aquatic systems should also take these sources into account.

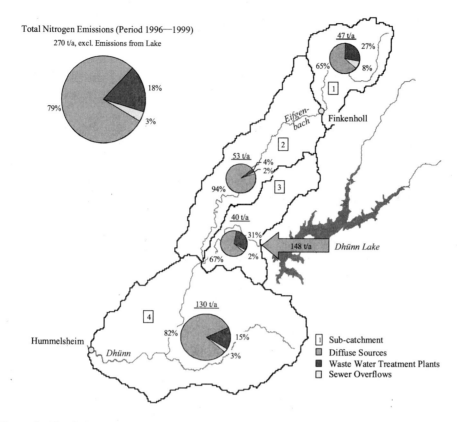

Figure 6 Total nitrogen emissions into the investigated catchment area for the period 1996—1999 assuming a deposition value of 25 kg/hm²·a (Nafo et Geiger, 2004b)

4.3 Ammonia toxicity in receiving waters

For the assessment of ammonia toxicity diffuse and background ammonia loads are assumed to be about 0.3 mg NH_4^+-N/L, pH of the surface water is 8 and temperature is 20°C According to the simple method. With this impact matrix the limit value of 0.1 mg NH_3-N/L is attained nowhere in the receiving water as consequence of sewer overflows and WWTP outflows of the two cities (Figure 7, left). Ammonia toxicity only occurs when pH in the surface water is 8.5. The highest toxic ammonia concentrations are calculated when sewer overflow volume is low due to poor mixing ratio of run-off volume and sewerage (Figure 7, right). Assuming the above mentioned values of diffuse and background ammonia loads, temperature of 20°C and pH 8 or 8.5 the limit value of 0.1 mg NH_3-N/L is attained nowhere in the downstream part of the river as consequence of point source discharges.

4.4 Dynamic of diffuse ammonia loads

To use physically based modelling, model calibration is a necessary step. For the assessment of the dynamics of diffuse ammonia emissions into the Dhünn river basin the SWAT model was therefore calibrated and validated with resulting good modelling efficiency in view of the water and mass balances in the river basin.

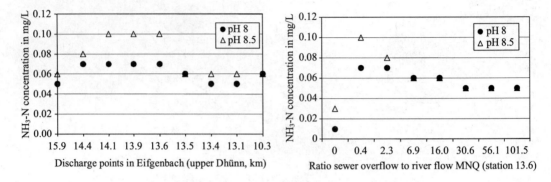

Figure 7 Calculated ammonia concentration in different drainage discharge points (left) and at station 13.6 in the upper Dhünn river (right) (Nafo et Geiger, 2004b)

For a better assessment of the significance of diffuse ammonia emissions into the river basin, simulated concentration duration curves are shown for different locations in Figure 8. The figure shows also decreasing ammonia concentrations in the lower Dhünnriver in comparison to the river concentration at confluence Eifgenbach-Dhünn. This is due to dilution effects caused by flow contribution of the Dhünnlake. This evidence indicates the influence of the existing lake on the water quality of the lower Dhünnriver. The concentration duration curves in Figure 8.indicates that the highest mean daily ammonia concentrations in dry weather periods only occur in few days throughout the year. It indicates also that the contribution of diffuse sources to mean daily ammonia emissions into the investigated river basin (maximum of 0.11 mg NH_4^+-N/L) is negligible regarding the assessment of ammonia toxicity in the receiving waters.

Without the dilution effect of the lake ammonia concentration in the Dhünn river downstream of confluence Eifgenbach-Dhünnduring dry weather periods can be higher than 0.3 mg NH_4^+-N/L (Figure 8). It can be concluded that existing point pollution sources in the upper area of the catchment should be taken into account to assess ammonia toxicity in the downstream receiving water when Dhünn lake were inexistent. Due to the fact that ammonia loss through nitrification in the upper Dhünn river (i.e. Eifgenbach) is negligible (see Niemann, 2001), the results indicate also that the emissions in the whole basin area should be analysed in order to define the required system boundary for ammonia toxicity assessment in downstream receiving waters.

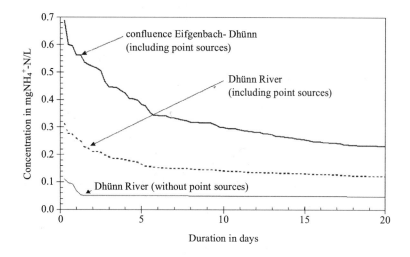

Figure 8   Simulated ammonia concentration duration in the Dhünn river at different locations for the period 1996—1999 (Nafo et Geiger, 2004b)

5   Conclusions

The contribution attempted to explain the main deficiencies of present water management practices and to show possibilities of accelerating the planning process by combining existing experience with newest developments such as remote sensing. The procedures recommended allow to specify cost effective measures and to assess these for their effects on receiving waters. Furthermore the recommended procedure also can be applied to all levels of conventional water management.

Water scarcity, pollution and flooding problems in river catchment and especially with large urban agglomerations prove that conventional approaches for water supply, flood control and pollution prevention have reached their limits. Alternative methods of water supply and drainage are available today. In a demonstration project in Beijing, China it has been proven, that measures combining latest water saving technologies, on-site storm runoff storage, treatment and infiltration can alleviate the pressure on water resources.

The on-going development in agriculture and urbanisation require repair measures in any case. Therefore the time is good, to rethink conventional approaches and introduce new technologies. However, the interaction of industrialisation, economics and environmental conditions as shown in Figure 1 usually only can be achieved in post-industrial societies. Still it is recommended to consider to implement the newest findings today rather than to copy systems of the past.

References

[1] ARNOLD, J.G., J.R. WILLIAMS, A.D. NICKS, N.B. SAMMONS. 1990. SWRRB: A Basin Scale Simulation Model for Soil and Water Resources Management. College Station, TX: TexasA & MUniversity Press.

[2] BEHRENDT H., HUBER P., KORNMILCH M, OPITZ D., SCHMOLL O., SCHOLZ G. and UEBE R. (2000). Nutrient Emissions into river basins of Germany. UBA-Texte 23/00.

[3] BWK-M3 (2001). Ableitung von Anforderungen an Niederschlagswassereinleitungen unter Berücksichtigung örtlicher Verhältnisse - BWK Merkblatt 3. Bund der Ingenieure für Wasserwirtschaft, Abfallwirtschaft und Kulturbau e.V., Düsseldorf.

[4] EMERSON K., RUSSO R.C., LUND R.E. and THURSTON R.V. (1975). Aqueous ammonia equilibrium calculations: Effect of pH and temperature. J. Fish. Res. Board Can. 32(12), 2379-2383.

[5] EMSCHERGENOSSENSCHAFT (1993), Materialien zum Umbau des Emscher-Systems, Wohin mit dem Regenwasser, Heft 7, Essen, January.

[6] EUROPEAN UNION (2000): Directive 2000/60/EC of the European Parliament and of the Council establishing a framework for the Community action in the field of water policy – The EU Water Framework Directive (WFD).

[7] GEIGER W. F., DREISEITL H. (2000), Neue Wege für das Regenwasser, Oldenbourg-Verlag, München, 2nd edition.

[8] HARTIG J. H., VALLENTYNE J.R. (1989), Ambio. XVIII(8), 423-428.

[9] NAFO I. I., GEIGER W. F. (2004a), The effects of spatial scales on the assessment of urban sources of pollution by implementing the EU WFD – Modelling requirements at regional scale, UDM 2004 Dresden (accepted for oral presentation).

[10] NAFO I.I., GEIGER W. F. (2004b), Assessment of the significance of nitrogen loads from point and diffuse sources in a small river basin, IWA Conference Marrakesch 2004 (accepted for oral presentation, peer review).

[11] NAFO I. I. (2004), Bilanzierung zur Beurteilung von Niederschlagswassereinleitungen auf regionaler Ebene. Schriftenreihe Forum Siedlungswasserwirtschaft und Abfallwirtschaft der Universität Essen, Heft 23, Diss.

[12] NEITSCH S. L., ARNOLD J. G., KINIRY J. R., SRINIVASAN R., WILLIAMS J.R. (2002).Soil and Water Assessment Tool - User's Manual 2000. Grassland, Soil and Water Research Laboratory, Agricultural Research Service, Temple, Texas.

[13] NIEMANN A. (2001). Schädigung des hyporheischen Interstitials kleiner Fließgewässer durch Niederschlagswassereinleitungen. Schriftenreihe Forum Siedlungswasserwirtschaft und Abfallwirtschaft der Universität Essen, Heft 15, Diss.

[14] UNESCO (1987). Manual on Drainage in Urbanized Areas, Volume I: Planning and Design of Drainage Systems, Volume II: Data Collection and Analysis for Drainage Design, Vol. I+II, W. F. Geiger, J. Marsalek, W. J. Rawls, F. C. Zuidema.

[15] UNESCO (1995). Integrated Water Resources Management in Urban and Surrounding Areas-A

Contribution to the International Hydrological Program of UNESCO (in print).

[16] UNITED NATIONS (1994). Guidelines on Integrated Environmental Management in Countries in Transition. New York.

[17] WORLD COMMISSION ON ENVIRONMENT AND DEVELOPMENT (1987): Our Common Future (Brundtland Report), Oxford University Press.

Flood Control and Groundwater Recharge
– Joint Chinese-German Demonstration Project for Beijing

W. F. Geiger
University of Duisburg-Essen, Germany

1 Introduction

Water supply of large cities is strongly linked to nearby and distant water resources. Many cities cover their water demand by local groundwater extraction, water reservoirs and rivers. Overdraft of groundwater lowers groundwater tables and in coastal regions causes saltwater intrusion, if groundwater levels come to be lower than sea level.

Beijing is listed among the world top ten cities suffering water scarcity. Besides water shortage, Beijing also faces local floods. 80% of the total annual rainfall of 640 mm appears within three months. High intensity storms during the rainy season result in immediate runoff from the mostly impervious surfaces not allowing for groundwater recharge. Figure 1 compares the seasonal rainfall depths of Beijing with Essen, Germany which represents middle European conditions.

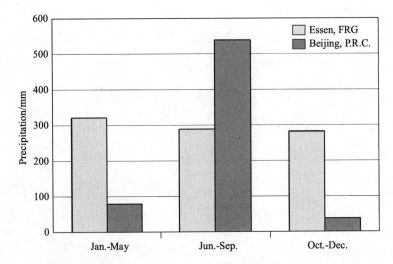

Figure 1 Comparison of average seasonal rainfall depths, Beijing with Essen, Germany

In Beijing following four categories of urban development may be distinguished that determine the techniques to be applied for flood control and groundwater recharge:
- Fully developed urban areas that remain unchanged (type 1)
- Previously developed urban areas that are completely redeveloped (type 2)
- Developed urban areas that are partially redeveloped or further condensed (type 3)
- Agricultural land, that is urbanised (type 4)

After thorough inspection five project areas were selected, which reflect above types. Some design details for three of the five project areas are explained, namely treatment of polluted storm runoff within the Institute of Geological Engineering (type 1), groundwater recharge in the new development of Tianxiu Garden (types 2 and 4) as well as water saving installations in the area of Engineering Foundation Works (type 3).

2 Hydrologic and hydro-geologic planning conditions

For any water system design aside from land use characteristics, climatological, hydrological and hydr-ogeological conditions must be assessed during the planning process. In case of drainage concepts that link flood control and groundwater recharge, rough or wrong assumptions easily may lead to over-design (waste of money) or under-design (increased risk of damages) of the systems. Therefore for each individual application hydrological and hydro-geological conditions must be evaluated in specific.

2.1 Hydrological conditions and derivation of design storms

Hydrologic data were available at seven meteorological stations. As most demonstration sites are within the vicinity of the Haidian station mainly data from this station were used. For derivation of design storms Shijingshan station was considered for its long and complete rainfall records (1981 to 1999). This station in 1994 showed the highest annual rainfall depth of 775.2 mm with a potential evaporation of 1 964.9 mm and the lowest of 333.0 mm with a potential evaporation of 2 154.5 mm recorded in 1997. The average annual rainfall was 553.0 mm. The average pan evaporation at same station was 1 826.1 mm. From the 19 years rainfall record duration-intensity dependencies and design storm patterns were derived, which were used for dimensioning of the storm water collection and storage system (Figure 2).

2.2 Hydrogeological and geological conditions

In Tianxiu Garden for subsoil characterization only 15 boreholes originally were available. In 2000 a number of 18 additional boreholes were drilled in Tianxiu Garden and the Engineering foundation Works areas. Most of these boreholes reached to a depth of 8 m, one was deeper. As from the borehole measurements differing up to 1.5 m the thickness of the sand and silt top layer was difficult to define and as most boreholes concentrated in the centre part of Tianxiu Garden, the measurements had to be extrapolated. WASY modelled the situation to define the lower level of the top layer using the TIN (Triangular Irregular Network) approach in combination with the Arc View-extension 3 D-analyst. Figure 3 shows the topography and Figure 4 the sand layer formation derived with the same modelling techniques.

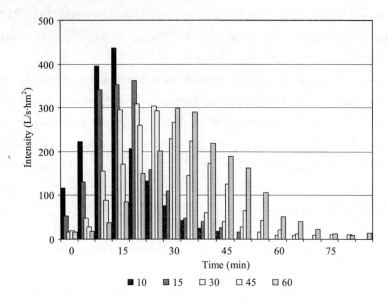

Figure 2 Design storm patterns used for sizing of different drainage components
(DORSCH CONSULT in BMBF, 2002)

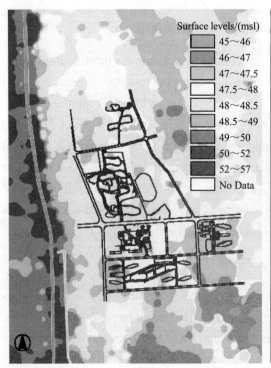

Figure 3 Digital Terrain Model including planned
Tianxiu Garden (WASY in BMBF, 2002)

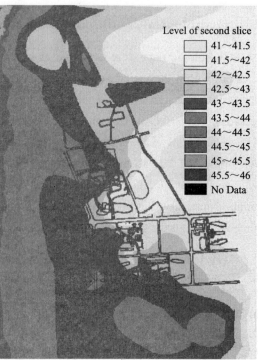

Figure 4 Levels of second slice
(WASY in BMBF, 2002)

The underground in Tianxiu Garden has an impervious top layer of varying depth followed by a second pervious fine sand layer. Beneath there is a compacted thin semi-impervious sand layer followed by pervious coarse sand.

To assess groundwater levels moreover monitoring stations in broader surrounding area were used. Again groundwater measurements were too rare for planning. Therefore WASY used the model FEFLOW (Diersch, 1998) to combine and complete the existing information to create a complete and detailed analysis of the groundwater situation.

It was found that for the small area of Engineering Foundation Works the existing profiles were sufficient for planning purposes, while for Tianxiu Garden available data were complemented by diverse simulation techniques. This remarkable effort is necessary to reduce the design risk. Furthermore, the modelling techniques allowed forming a groundwater model which later also helped to assess the effect of groundwater recharge at the sites.

3 Example for a storm water treatment facility

Storm water treatment facilities were constructed in three of the project sites. Discussed in detail is a facility which treated polluted runoff in the Institute of Geological Engineering. The yard surface of around 9.200 m^2 connected to the structure is used for parking. The water drains to gutters covered by grids. The peak inflow at a 5-min storm of a return period of 5 years was calculated to 120 L/s. The planning goal was, that as much storm water as possible shall be recharged into the ground. No excess runoff should occur up to a 5-year storm. The tank structure is emptied into two recharge wells which are able to take a maximum flow of around 4 L/s each, 8 L/s in total. The system was dimensioned by hydrodynamic runoff simulations using the model HYDROCAD of Dorsch Consult.

The structure shown in Figure 5 consists of a rectangular covered basin with a width of 6.50 m and a length of around 23.25 m (exterior dimensions, including walls). The variation in water level during regular operation is 2.90 m which yields a storage volume of around 391 m^3 in the tank. This corresponds to a specific volume of 425 m^3/hm^2 equivalent to 42.5 mm of rainfall which is stored in the structure. Additional storage is activated within the collection system during storms.

3.1 First Flush Cell

The initial phase of storm water runoff is particularly polluted. This phenomenon is referred to as first flush. This extremely polluted portion of runoff should be discharged into the sanitary sewer. For this task a first flush cell is provided. It has a volume of 18 m^3, which equals approximately 2 mm of rainfall from the impermeable surface of 9.200 m^2.

After each rain event, the first flush cell is emptied by a pump. In order to protect the pump and to keep gross solids away from the sanitary sewer, the sludge at the bottom of the cell remains in the cell. This is why after each storm event the first flush cell should be inspected and, if necessary, cleaned by a suction truck or by hand.

196 Cost-Effective Measures for Flood Control and Groundwater Recharge in Coastal Areas

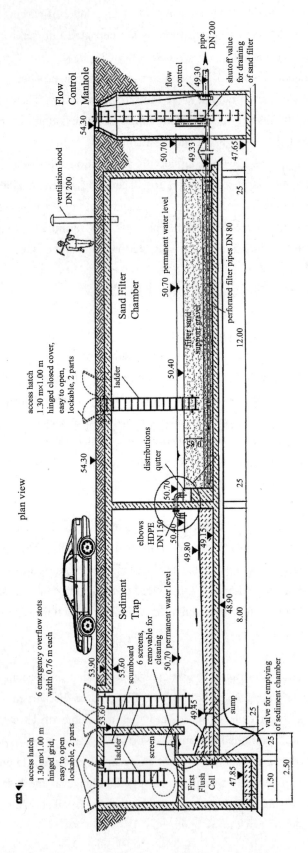

Figure 5   Plan view of the storm water treatment facility (UFT in BMBF, 2001a)

3.2 Screen Compartment

As soon as the first flush cell is filled up, the water entering the structure is fed over a low side weir onto 6 removable static screens (stainless steel punched metal with 8 mm holes, size 750 mm × 750 mm) of a total area of 3.37 m². The openings are arranged slightly above the permanent water level to allow falling-dry and to prevent fouling of collected organic material. As soon as the water level has risen slightly, however, they will be immersed fully in the water to ensure minimum surface loading. In case of clogging, water may overflow into the tank over openings in the wall close to the ceiling. The screen compartment acts also as an oil trap. Nevertheless, it is acceptable that some oil may escape into the next chamber through the overflow opening, e.g. if the tank is completely filled.

The screens need to be inspected regularly and, if necessary, cleaned by hand. Therefore they can easily be removed manually with two handlebars.

3.3 Sediment Trap

The tank has a permanent water level of about 1 m. Sediments in the storm water which pass the screens will settle down in the sediment trap chamber. Six elbow overflow pipes form the overflow to the next chamber. This arrangement captures remaining traces of oil in this chamber. At an outflow of $Q = 8$ L/s, the surface loading of this chamber is as low as $qA = 0.60$ m/h. This ensures good sediment removal. From time to time, the chamber should be emptied entirely including the accumulated sediment sludge by an outlet pipe which leads to the first flush cell from where it is pumped into the sanitary sewer.

3.4 Sand Filter Chamber

The last compartment features a bottom sand layer of 1.25m thickness (1.00 m over the top of the filter pipes). Water is infiltrating from above to the bottom where it is drained by perforated filter pipes leading to the outlet. The surface loading of the sand filter at a flow of $Q = 8$ L/s is $qA = 0.40$ m/h which is a maximum for this type of sand filter.

The sand filter consists of two layers of different grain sizes. The upper, finer sand is the active filter material while the lower, somewhat coarser layer is supporting the fine sand and preventing it from being entrained into the filter pipes. The filter pipe system has been selected with regard to the used material for the support layer which is embedding the pipes. The stainless-steel pipes have a diameter of DN 80.

The sand filter chamber has a large access opening in case the filter material has to be replaced after several years of operation or other kind of maintenance work is necessary. Since the chamber should remain dark to prevent excessive growth of algae, only solid covers should be used for the opening.

3.5 Flow Control Manhole

The filter pipes are fed to a separate manhole where the off going pipe to the recharge wells is at the level of the permanent water level. The bottom of this manhole is at a somewhat deeper level. If necessary, water from this manhole can be pumped out to the recharge well with a mobile submerged pump; the filter is drained by this operation. Usually, the outflow is limited by the recharge capacity of the wells. If desired, the outflow to the recharge well can be equipped with an extra flow control unit, though. By measuring water level differences in the manhole and in the recharge wells, flow can be monitored.

4 Example for planning and design of groundwater recharge facilities

Groundwater recharge facilities were installed in all project areas. Explained is an example of one of the two facilities which were built in Tianxiu Garden. The objectives are to harvest and infiltrate runoff from roofs, streets and yards preventing inundating. Thus the local aquifer is augmented by infiltration of runoff into the underground. As explained above runoff from roofs is collected in storage sewers and pipelines and treated before it reaches the recharge wells. Furthermore, for retention purposes a part of the roof system is directly connected to storage tanks, which are used to balance evaporation losses in an artificial lake.

4.1 Specifications of the storage tanks

The storage volume needed to balance evaporation losses in the artificial lake depends on evaporation rates and, especially, on duration of periods with poor precipitation. Due to strong fluctuations of the average monthly evaporation in Beijing, different typical evaporation values are selected to calculate the required storage volume. Figure 6 presents a graphical solution for determination of the required storage tank volume based on the lake surface of 4 260 m^2.

For instance, a typical dry period of 60 days with an average evaporation rate of 80 mm per month demands a storage volume of at least 682 m^3. In the dry season, rain events will be rare and also rather reduced. Therefore, it cannot be expected that the storage volume will be refilled. Thus, even a very large storage tank would not be able to balance evaporation losses in the dry season completely. Moreover, the annual volume of water which is recharged into the groundwater would be considerably reduced because most of the harvested water is used for alimentation of the artificial lake.

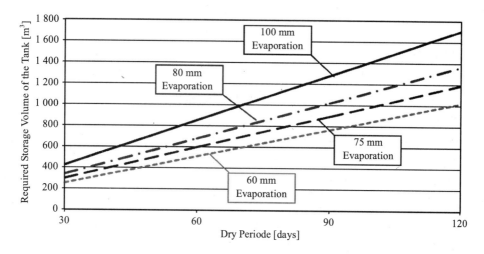

Figure 6 Required storage volume of the tank for the artificial lake

4.2 Major structures of the roof drainage system

The water in the roof drainage system is expected to be feebly polluted mainly with dust and pollen. Before the water is fed into recharge wells, one treatment structure is planned for each of both recharge facilities. The treatment structure serve the following functions:

- Capturing of the polluted first flush at the beginning of storm runoff after long dry periods
- Dust removal, trap of floatable materials
- Handing over water to the groundwater recharge wells

The structures have a first flush chamber which is filled by the first 2 mm of precipitation of each storm event. The areas connected to the northern and southern roof drainage system are nearly equal with 1.17 and 1.12 hm^2, respectively. The volume of the first flush cells is 21.8 m^3 for each structure which is pumped to the sanitary sewer. After the first flush chamber has been filled, the water is fed to each group of recharge wells by a single DN 200 pipe. The recharge wells finally will fill up to the same water level like in the sewers. The discharge of the recharge wells is limited by the hydraulic capacity of the underground. The four recharge wells at each site are connected by single DN 200 pipes without further valves. After complete emptying of the structure and also of the sewer system to the recharge well, the first flush cell is emptied finally to the sanitary sewer.

This is done by a small submerged pump (Q = 1 to 2 L/s). It must be capable of pumping strongly polluted water and also sand. In order to maximize the amount of water available for groundwater recharge, the pump should not run too frequently or too long. The pump control should be operating automatically, depending on the measured water level in the structure. At regular time intervals, the structure must be inspected. Remaining sediments in the first flush chamber should be removed manually or by a vacuum suction truck.

4.3 Specifications of the recharge wells

The main purpose of the wells is to guarantee a large volume of groundwater recharge. Therefore two identical infiltration facilities exist, one in the North and one in the South of Tianxiu Garden. Each facility consists of four recharge wells. The basis for design are hydrogeological studies, which imply water tables, permeability, slice elevations and boundary conditions. The diameters of the recharge wells are 2m, with shafts made of concrete rings. Water tables of September 1998 have been used as the decisive water level for calculating infiltration capacities. September 1998 is the months with the highest water tables observed so far.

As the return period for the infiltration times cannot be calculated in advance, several alternative situations have been simulated. The simulations represent ground water tables with return periods of 10, 5, 2 and 1 year(s) and water levels within the wells of 47.5, 47, 46, 45 and 44.5 m (msl). From these results it can be concluded, that based on the 11-year period of groundwaterdata at Xiaojiahe only once in ten years infiltration capacities of 5.7 L/s (South) and 4.7 L/s (North) cannot be guaranteed between June and September. Once in 5 years this critical capacity amounts to 6.7 L/s (South) and 5.4 L/s (North). Both figures are based on a conceivable water level within the wells of 47.5 m above msl. Figure 7 displays FEFLOW-model results showing the difference between uninfluenced ground water and the ground water at the end of the infiltration period (0.5 days, southern facility).

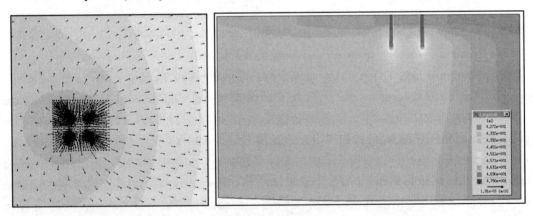

Figure 7 Flow direction during infiltration directly below the well (left side)
West-east cross-section during infiltration showing heads (right side) (WASY in BMBF, 2002)

The distance (center to center) between two neighboring wells in one facility was determined to be at least 17.5 m in order to avoid too much interference between both wells. Such an interaction would reduce the total recharge capacity. The distance between a well and the surrounding basements normally amounts to 1.5 times the depth of excavation, so that the wells should be placed at least 12 m from the nearest building. For construction of the wells concrete rings were chosen with heights of 0.5 and 1m. The southern facility has a total depth of

9.55m for all wells and the northern facility has a total depth of 7.75m.

The infiltration predominantly takes place in a vertical direction at the shaft bottom. Since the refill of the excavation will be carried out with non-cohesive, permeable material, an additional small infiltration could be integrated along the perforated pipes (DN 200) which connect all wells within one facility. This additional infiltration has not been taken into account within the simulations mentioned. The shaft bottom is filled with an approximately 1.0 m thick coarse gravel layer, which is covered by a stone riprap as a protection against erosion. These layers can be exchanged if necessary. The gravel layer has to have a permeability higher than 5×10^{-3} m/s.

5 Role of domestic water saving for sustainable urban water concepts

The first concern is awareness of people about the importance of water-saving and the true costs of water. Technical water saving schemes including rainwater harvesting and greywater recycling were realized in two residential buildings. In semi-arid regions with low rainfall and high population density as it is the case in Beijing recycling of slightly polluted domestic wastewater (greywater) for reuse to flush toilets is an effective measure to reduce potable water consumption. The aim was, to show that the combination of greywater recycling and rainwater harvesting can be an essential part of water saving. The greywater recycling and domestic rainwater harvesting system was designed and installed by GEP.

5.1 Concept of the greywater recycling plant

The selected residential buildings in the area of Engineering Foundation Works were newly constructed. They have six stories each with ten residential units per level. The middle to upper level apartments are sold after construction.

Figure 8 provides a plan view of the two buildings showing the location of rainwater, greywater and blackwater lines. The greywater recycling facility was placed in between the two buildings.

In order to show that the reduction in water consumption in a household is possible without any of loss of comfort, building no. 6 was fitted with water-saving devices. In building 7 standard fittings were used. In both buildings toilets were equipped with economical 6 liters–flushing boxes.

The roof drainage of both buildings was collected and stored after filtering beneath the greywater demonstration center. The greywater from showers, wash basins and baths from building 6 are processed and reused for flushing toilets. The sludge arising during biological processing is directed to a septic tank, which also treats the blackwater from the two buildings. The overflow from the rainwater storages enters the wastewater drain. The overall concept of the demonstration plant is shown in Figure 9. For filtering a slotted strainer device from GEP was used.

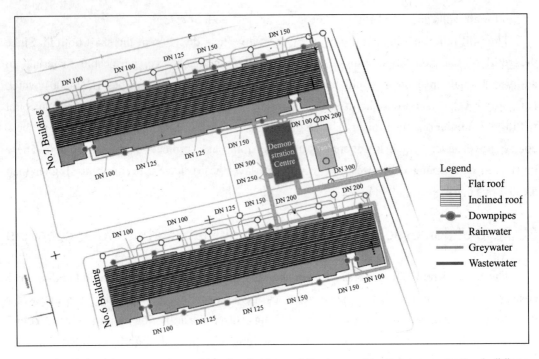

Figure 8 Rainwater, greywater and blackwater lines of the two residential demonstration buildings in Engineering Foundation Works (GEP in BMBF, 2001c)

Both the rainwater harvesting system and the greywater system were equipped with control and regulatory technology to measure the fluxes and the degree of filling. Remote monitoring and maintenance now already has been done for several years.

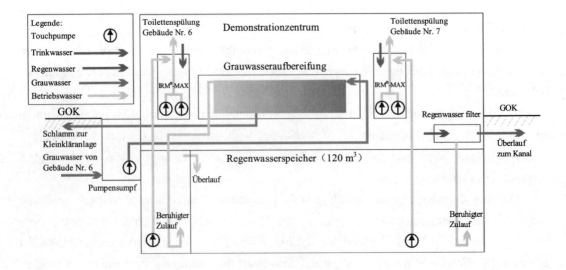

Figure 9 Schematic of installations in the rainwater harvesting and greywater-recycling center in Engineering Foundation Works

5.2 Dimensioning of the facilities

The dimensioning of the storage volume usually is based on annual rainfall. However, as already Figure 1 indicated, the unequal distribution of rainfall has to be taken into account. Further the proportion of flat and slope roofs had to be considered. This resulted in a usable storage of 668 m^3 of roof drainage. Accordingly, only 29% of the required water for toilet flushing can be covered by rainwater.

The simulation based on the daily rainfall measurements at the Haidian weather station from 1981 to 1993 yielded, that with a storage volume of 120 m^3 90% of the roof drainage water can be utilized. Then the storage would overflow 30 times in 14 years. However, the rainwater storage tank due to hydrologic conditions would be more or less empty from mid-October to mid-April.

The greywater recycling plant was designed for a daily volume of 10.000 liters. The calculated requirement for flushing toilets, however, only would be 6.000 liters per day. Figure 10 provides a view of the greywater recycling plant in Engineering Foundation Works.

Figure 10   Greywater recycling facility in Engineering Foundation Works

It should be noted that greywater is independent of seasonal variations in water volume and is therefore a most suitable technique for water-saving especially under semi-arid conditions. However, compared to rainwater utilization, the technology involved is much more complex and therefore higher investment and operational costs incur.

6 Conclusions

When designing water sensitive concepts especially under semi-arid conditions a combination of rainwater harvesting from as many surfaces as possible should be considered, both for groundwater recharge, household and commercial uses, coupled with greywater recycling of used water which is less polluted. This will yield the largest potential for long-term saving of potable water especially in residential areas. Installation of fully decentralized and independent systems, i. e. using composting toilets, the ECOSAN (Lange and Otterpohl, 2000) system etc. still has to be proven that they can be safely applied to large urban areas. However, these systems seem to be quite suitable for rural cities up to a population of about 50.000 inhabitants.

In any case, thorough analysis of hydrological and geological data is necessary in order to avoid operational risks and false investments. It also is advisable to use in densely populated area advanced planning tools such as HYDROCAD and FEFLOW. Application of these tools force to thoroughly analyze the local situation, but finally helps to find a suitable solution.

Last not least public awareness plays a dominant role in operating sustainable water concepts. Public awareness also raises understanding for costs involved. Of course in any case the user should pay for the real cost of water.

References

[1] BMBF (2001a): Drainage Concept for institute of geological engineering in Bali Zhuang, Hai Dian District, Beijing, Final Technical Report, to be released in 2004.

[2] BMBF (2001b): Drainage Concept for Demonstration Site II Beijing – Engineering Foundation Works, Part EFW, Preplanning Report, to be released in 2004.

[3] BMBF (2001c): Demonstration Center for Rainwater Harvesting and Greywater Recycling in Demonstration Site II – Part EFW, Technical Report by GEP, to be released in 2004.

[4] BMBF (2002): Drainage Concept for Demonstration Site II - Tianxiu Garden, Part B, Final Technical Report, to be released in 2004.

[5] DIERSCH, H. J.-D., (1998): Grundwassersimulationssystem FEFLOW. User-/Reference-Manual. WASY GmbH, Berlin, 1998.

[6] FEFLOW, Produced by WASY GmbH, Berlin, www.wasy.de.

[7] GEIGER, W., LI, Z., (2002): Water and mass balances in urban areas, to be released in 2004.

[8] HYDROCAD (2003): Stormwater Modeling System, www.hydrocad.net.

[9] KOENIGER, W. (1981): The Application of the Extreme-Value-Type 3 Distribution for Rainfall and Low Flow Analysis. gwf - *wasser/abwasser,* No. 10, pp. 460-466.

[10] KOENIGER, W., GRAF, M. (1996): Modelling of Complex Hydraulic Structures in Urban Drainage Systems by a Generalised Rainfall-Runoff Model. Paper presented at the 7th *Int. Conference on Urban Storm Drainage,* Hannover, September 9-13, 1996.

[11] LANGE, J., OTTERPOHL, R. (2000): ABWASSER- Handbuch zu einer zukunftsfähigen Wasserwirtschaft. MALLBETON-Eigenverlag.

[12] OTTER, J., KOENIGER, W. (1986): Design Storms for Sewer Networks, Overflow Structures and Retention Basins. *Gas – Wasser – Abwasser,* No. 3, pp. 124-128.

Engineering View of Cost Efficiency

W. F. Geiger   P. Meyer
University of Duisburg-Essen, Germany

1 The necessity of identifying cost efficient measures

Communities, industries and diffuse sources, usually resulting from agricultural runoff, mining etc. mainly affect water resources. Resulting water quality problems such as temperature increase, dissolved oxygen deficit and eutrophication usually define the status of receiving waters. Measures have to be identified and implemented at different scales: on river catchment level and on water body. Deficits, planning conditions and priorities may differ according to scale.

For a cost effective implementation of measures it is necessary to develop a clear perception on useful measures, their efficiency and costs. Identification of measures must take specific local and regional conditions into consideration often in very detail. Therefore the finally implemented measures may differ from general measure collections. In order to keep this difference as small as possible, it is necessary to develop a conclusive procedure that suits all scales.

This actually is required in the central document of the European Water politics, the European Water Framework Directive (EUROPÄISCHE UNION, 2000) in form of operational plans for the management of river catchments. These operational plans are essential instruments, which need not only list the measures chosen for conservation or improvement of the good status of waters but also economic functions to ensure the implementation of cost effective measures and their on-going maintenance.

2 Consecutive steps to achieve cost efficiency

It requires aneutral procedure to define cost effective measures for deficits in a catchment. Potential measures are measures, which can be installed at an actual source in a catchment. The local circumstances must allow implementation and must ensure that it has no negative impacts on the water body/catchment. It would be useful to have catalogues of all possible measures available to choose the most suitable ones to close the deficits. Such catalogues (tool boxes) would have a general validity and could be applied in various projects. In this tool boxes the different measures must be described according to efficiencies and costs. The procedure suggested in Figure 1 shall be trailed to gain this cost efficiency.

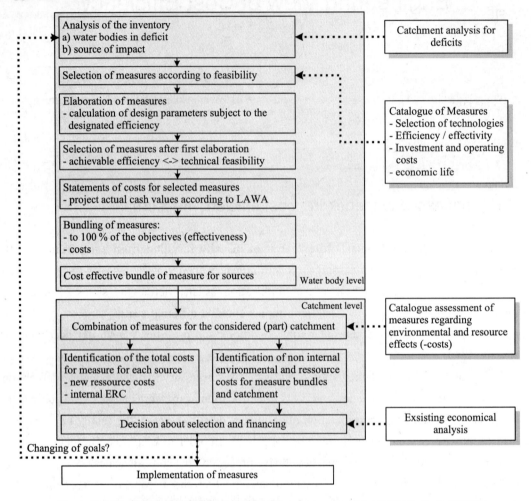

Figure 1   Working steps to gain cost effective measures (LONDONG et al., 2005)

The Ministry for Environment, Conservation, Agriculture and Consumerism of North Rhine-Westphalia (MUNLV), Germany asked the University of Duisburg-Essen, the Bauhaus University of Weimar and the Ruhr-Research Institute for Innovation and Structural Policy (RUFIS), Bochum to develop and test the method described to derive cost effective measures.

For a better understanding the problem to define the efficiencies of measures will be discussed first, then different costs will be defined and the combination of measures and the evaluation of the cost efficiency will be described. It will be discussed first in the context with achieving good ecological and chemical status in water bodies.

2.1   Definition of measure efficiency

The necessary efficiency of a measure in a specific application depends on the status, which has to be achieved for a good ecological and chemical status of water bodies and on the inflow concentration of the point source. Therefore the point source has to be identified and classified and the potential measures should be selected from the measure catalogue (Figure 2). Afterwards the efficiency of the selected measures has to be examined for the local situation. It is suggested to express the efficiency in 10%-steps and relate the 100% to the difference between the requested situation in the water body and the actual situation. In consequence the possibility is given for an easy combination of measure efficiencies according to the objectives at the point source, where about the reached objective is equal to 100% efficiency. According to formulas given in the measure catalogue the efficiency of feasible measures should be calculated. For instance, given a required BOD level of 6 mg/L in the water body and a point source discharging 18 mg/L with a one third flow rate of the receiving water with a concentration of 3 mg/L the discharge of 18 mg/L leads to an average concentration of 9 mg/L in the water body. Applying a measure for treatment which reduces the inflow concentration from 18 mg/L to 9 mg/L results in the efficiency of 100%.

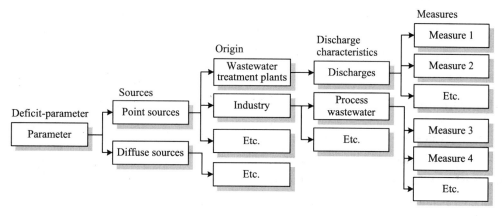

Figure 2   General procedure to identify measures according to the source (GEIGER et al., 2005)

2.2 Definition of costs

The different costs to evaluate a cost effective measures are the investment and operation costs and supplementary environmental and resource costs (ERC). The investment costs are the costs for constructing a measure. The operational costs are the costs for operation and maintenance of this measure during its lifespan. Environmental costs are defined as the costs of damage that water uses impose on the environment and ecosystems and those who use the environment (e.g. a reduction in the ecological quality of aquatic ecosystems or the salinization and degradation of productive soils). Resource costs are defined as the costs of foregone opportunities which other uses suffer due to the depletion of the resource beyond its natural rate of recharge or recovery (e.g. linked to the over abstraction of groundwater). Investment and operational costs mostly are private, while ERC often are both private and public costs (WATECO, 2002).

The investment and operational costs of different potential measures for achieving a better water status have to be calculated for a certain lifespan on the basis of cost comparison methods. In each year a price increase and a discount rate has to be considered to calculate the actual cash values. On this basis the costs of the different measures can be compared. Additionally environmental and resource costs should be added to receive a cost analyses including ERC.

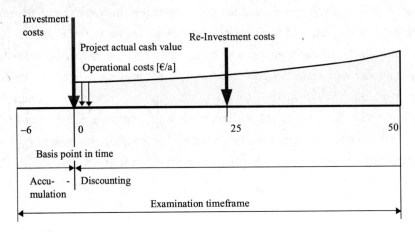

Figure 3   Calculating the project actual cash values of different measures (LONDONG et al., 2005)

2.3   Evaluation of cost efficiency

After evaluating the costs and the efficiency of the different measures according to the above-described methods, these measures have to be combined by minimizing costs and maximizing effects. First of all various measures with a total efficiency of 100% for one considered parameter have to be determined. Afterwards the efficiency of all measures of this bundle should be combined to get all possible measure bundles. Hence the efficiency for the next considered parameter is to be calculated for the chosen measure bundles. If that efficiency does not reach 100%, further measures or measure bundles according to the same procedure have to be considered until the efficiency of all deficit parameters are combined up to 100%. For each single measure the actual cash value is determined in relation to the volitional efficiency. Adding each single project actual cash value derives the amount of the actual cash values of the whole combination. At this point the different bundles of measures reach the same efficiency but may have different costs. The most cost efficient measure respectively measure combination is found by reaching the smallest amount of project actual cash values. How the efficiency and costs of different measures regarding different parameters are combined is shown in the flowchart in Figure 4. First of all measures for the first parameter (parameter x) is selected from the catalogue of measures and for an efficiency of 100% the related costs are calculated. Afterwards the efficiency for the second considered parameter (parameter y) of this measure I is calculated for the second parameter (parameter y). Now additional measures are selected from the tool box for parameter y to reach an efficiency of 100% and therefore the related costs are calculated.

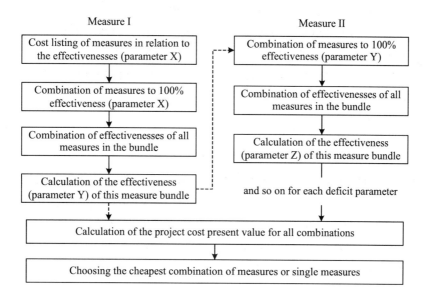

Figure 4   Evaluation of cost effective measures (GEIGER et al., 2005)

3   Consideration of water bodies or catchments

Water bodies can be characterized by hydrological, physic-chemical, biological and hydro-morphological parameters. In the following chapters we will discuss whether measures, their effects and related costs have to be identified for individual water bodies and/or river catchments as a whole.

3.1   Necessity for consideration of individual water bodies

To reach a good ecological and chemical status it is necessary to analyse every single water body. During analysing the water bodies all different pollution sources have to be identified. After defining the pollution sources it is important to distinguish all different sources in primary sources and non-primary sources. The primary sources influence the main river (whole downstream catchment) while the secondary sources have consequences for on local water bodies only. For each single situation it has to be decided whether a measure on a secondary source should be implemented or whether measures at primary sources suffice to reduce pollutants. It might be an option, not to gain a good ecological and chemical status in a small creek flowing into a large river. For evaluating whether a source is a primary or a secondary source it is important to analyse every single water body.

3.2   Presentation on river catchment scale

The river catchment scale has to be analysed in any case. This condition deals with the input of downstream water bodies. In a catchment only the detailed analysis according to 3.1 will allow the decision whether all measures shall be applied or whether some water bodies can remain in their inadequate status. In other words before stating which measure should be applied it has to be decided whether every single water body has to be in a good ecological and chemical status, or whether some smaller water bodies can violate the goal as long as the good status is achieved on the average in the whole catchment.

4 Example for the derivation of cost effective measures

This application was supposed to serve as an example for the many national environmental offices, which have to fulfil the WFD. The example shown was the Lippe River catchment (area 4 881.8 km^2) for the deficit parameters of temperature and chloride. A general catalogue of measure for temperature and chloride has been set up for calculation the efficiency and cost of a chosen measure in the local situation. An extract of this measure catalogue gives Table 1.

Table 1   Extract from the catalogue of measure (LONDONG et al., 2005)

No.	Measure	Efficiency grade (EG) ⇔ reference value	Costs
5.2	Usage of heat: Sewer – heat exchanger	Heat: $L_{Waermetauscher} = \dfrac{4186 \cdot 998 \cdot Q_E \cdot \Delta\vartheta_{W,Soll,Wi} \cdot ZEG_{Wi} \cdot \eta_{Waermetauscher}}{\approx 4.000 \text{ W/m}} \leqslant 300 \text{ m}$	Investment costs heat exchanger, heat pump + other components and construction costs: IC ≈ 35.126·e0.0151·L [€] without additional water supply net IC ≈ 128.777·e0.012·L [€] with water supply net Operational costs for pumps, Maintenance: OC ≈ 298.4·L + 5959 [€/a] Estimated benefit: OC$_{benefit}$ ≈ –2.5 €/(a·m^2 connected surface) Lifespan: Ca. 10 years pumps Ca. 50 years heat exchanger

First of all the measure for the parameters temperature and chloride the measures were selected from this catalogue of measure according to the point source. The following figure 5 gives an overview over all potential measures, which could be implemented at a specific source regarding the parameters temperature and chloride.

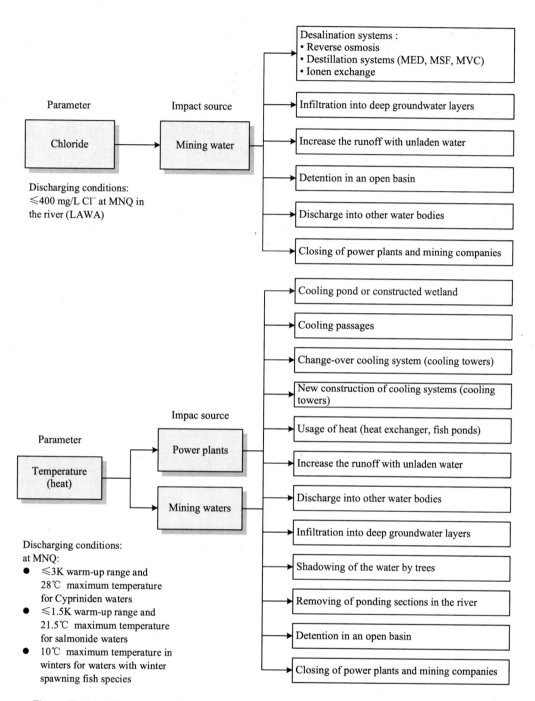

Figure 5   Identifying potential measures for reducing chloride and temperature problems (LONDONG et al., 2005)

For the Lippe River catchment the working steps according to Figure 1 were followed considering the parameters chloride and temperature.

4.1 Problem assessment and identification of sources

For the Lippe River catchment an assessment report was available (Stua Lippstadt, 2004). In this assessment report as goals for chloride locally and in the region a concentration of 400 mg/L is demanded, while at the border to the Netherlands the Rhine River concentration of chloride should not exceed 200 mg/L (BMU, 2004) (The Lippe River is flowing into the Rhine River and then the Rhine River is crossing the border). For temperature locally and regional as well as at the discharge point to the Rhine River during summer a temperature of 28°C (cyprinid waters) (21.5°C for salmon waters) and during winter a temperature of 10°C should not be exceeded. The maximum temperature increase should not exceed 3°K (cyprinid waters) (1.5°K for salmon waters) (78/659/EWG).

The sources for chloride were subdivided into point and diffuse sources. While diffuse sources are only due to geogenic conditions, point sources result from coal mining, industry, waste water treatment plants and combined sewer overflows. From coal mining drainage water is discharged into the Lippe River and its tributaries. Industries discharge cooling water and process water, in winter salt is discharged via rain and snow melting water from de-icing streets and yards. Sewage treatment plants receive waste water from industries including chemical loadings, which also contribute to the chloride in the discharge as well as de-icing in winter. Finally combined sewer overflows mainly contain de-icing residuals.

Temperature problems are caused by point sources only, which are coal mining, industry and sewer treatment plant effluents. The discharged salt drainage water from coal mining has a high temperature. Industry mainly discharges cooling water and warmed-up process waters. Figure 6 provides an overview of the most important sources for temperatures and chloride in the river catchment.

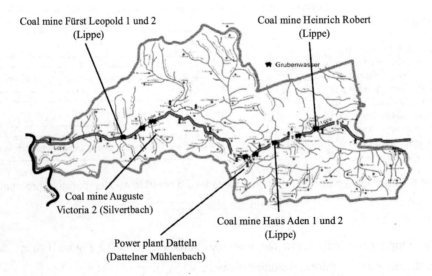

Figure 6 Sources for temperature and chloride in the Lippe River catchment
(LONDONG et al., 2005)

After analysing the sources for the deficits according to the inventory the technically feasible measures for the identified point sources were selected analogous to Figure 2. Afterwards the efficiencies and costs for these measures were elaborated.

4.2 Estimation of costs and selection of cost-effective measures

Precise costs could only be provided for investment and operation. Identification of environmental and resource costs for individual measures was nearly impossible. For individual measures only investment and operational costs can be estimated.

To give an example four cost-effective measures for the discharge of the coal mine Auguste Victoria 2 were analysed according to the method described above. The efficiencies and the associated costs were evaluated (Table 2).

Table 2 Examples for efficiencies and project actual cash values for measure at coal mine Auguste Victoria 2 (LONDONG et al., 2005)

	Measure bundles	Efficiencies	Project actual cash values
1	Redirection to the river Rhine (39 km)	100% chloride 100% temperature (SLC/WLC)	27.2 million €
2	Redirection to the Rhine	90% temperature (SLC) 80% temperature (WLC) 80% chloride	20.8 million €
	Unplanted pond	10% temperature (SLC) 40% temperature (WLC) 0% chloride	200.000 €
	Reverse osmosis	20% chloride	9.4 million €
		Total:	30.4 million €

SLC – summer loading case
WLC – winter loading case

For the whole Lippe River catchment several measure bundles reached the required good status for the whole Lippe River catchment. Two of the measure bundles in respect to their investment and operational costs actual cash values were 15.6 mil € and 16.7 mil €. However, both options include infiltration in deeper groundwater layers. If this is not possible, the next cheapest measure bundle amounts to 45.2 mil. €. It was impossible to include environmental and resource costs on the measure level, as these were simply not available for the required detail. Environmental and resource effects may be used to compare measure bundles, which show the same cost magnitude.

An indication for the magnitude of ERC on the catchment scale is reported by HOLM-MÜLLER et al. (1991) who conducted a survey of the willingness to pay on household basis for river improvement towards a better ecological status. Using the reported benefit-transfer rate a sum of 12 million € per year can be calculated for the Lippe River catchment. The different measures are connected with different environmental and resource effects. Most of them can be evaluated by market prices for energy and water. Only the evaluation of changes in landscape is difficult due to missing data.

Another important problem is the financing the selected measures. In the described procedure minimizing the sum of the costs has been the goal. However, measures have to be paid by polluters. Therefore another routine, which allows finding a fair distribution of costs, has to be found.

4.3 Adjustment of the practical procedure

Concluding the results of the project in the Lippe River catchment for an involvement of ERC it can be stated that it is impossible to gain sufficient data regarding the ERC. For measures affecting special deficit parameters the ERC are marginal. Only for examining the whole program of measure considering all measures, which have to be installed to achieve a good ecological and chemical status, the ERC can be estimated roughly. So for single measures only the environmental and resource effects can be involved to evaluate cost effective measures. On this basis a new sequence of working steps to achieve the required goals was developed. With step 6 the assessment of the environmental and resource effects shall be applied in place of the environmental and resource costs.

Table 3 Comparison of theoretical working steps and the modified ones after adaptation the project for achieving the goals of the WFD (GEIGER et al., 2005)

Step	Theoretical working steps	Practical working steps
1	Analysis of the assessment in view of problems and identification of pollution sources	Analysis of the inventory in view of problems and identification of pollution sources
2	Derivation of potential measures to cope with the problems	Derivation of potential measures to cope with the problems
3	Identification of measures suitable to achieve good quality status	Identification of measures suitable to achieve good quality status
4	Selection of measures after elaboration	Prove of technical feasibility of measures
5	Identification of costs for all measures and measure combinations including ERC	Identification of costs for all measures and measure combinations only on the basis of investment and operational costs
6	Selection of the cost-efficient measure bundle	Pre-selection of measure bundles
7	Financing options and final selection of measures or measure bundles	Assessment of bundles with scaling environmental and resource effects
8	Realization of measures	Financing options and final selection of measures or measure bundles
9	Control of their effects	Realization of measures
10		Control of their effects

EU-Wasserrahmenrichtlinie – Beispiel Lippe. Projekt gefördert durch das Ministerium für Umwelt, Naturschutz, Landwirtschaft und Verbraucherschutz des Landes Nordrhein-Westfalen. Aktenzeichen: IV-9-042 049.

[6] WATECO (2002): Economics and the Environment – The Implementation Challenge of the Water Framework Directive – A Guidance Document.

5 Conclusions and Recommendation

Until today the overall procedure to select cost effective measures is deriving measures or measure bundles according to expert knowledge. The selection of measures or measure bundles along with this procedure is dependent on this knowledge and therefore subjective. For reaching a unbiased selection of measures and/or measure bundles this detailed procedure described is an improved possibility to gain true cost efficiency.

The cost effective program of measure including ERC cannot be implemented including ERC for single parameters/measures. For determination of the ERC for a whole quality improvement of a river catchment basically ERC can be derived. With regards to the problem, that no real environmental and resource costs can be calculated for single measures/parameters the environmental and resource effects should be the starting point for evaluating a program of measures, but here only on the river catchment scale.

While assessing the different parameters and defining the measures a problem is to determine the cost effective program of measures by hand for more than two parameters. For more than twoparameters a computerized method must be considered.

Further criteria should be given at which complexity comprehensive quality modeling becomes necessary. Finally it should be investigated in as much index systems are suitable to replace cost calculations at all.

References

[1] 78/659/EWG: Richtlinie des Rates 78/659/EWG vom 18. Juli 1978 über die Qualität von Süßwasser, das schutz- oder verbesserungsbedürftig ist, um das Leben von Fischen zu erhalten (ABl. Nr. L 222 vom 14.8.1978 S. 1; Beitrittsakte Griechenland - ABl. Nr. L 291 vom 19.11. 1979 S. 17); Beitrittsakte Spanien, Portugal - ABl. Nr. L 302, 15.11. 1985 S. 9); 90/656/EWG - ABl. Nr. L 353 vom 17.12. 1990 S. 59); VO (EG) 807/2003 - (ABl. Nr. L 122 vom:: 16.5.2003 S. 36) gültig bis 22.12. 2013 gemäß Art. 22 RL 2000/60/EG.

[2] Europäische Union (2000): Richtlinie 2000/60/EG des Europäischen Parlaments und des Rates vom 23. Oktober 2000 zur Schaffung eines Ordnungsrahmens für Maßnahmen der Gemeinschaft im Bereich der Wasserpolitik. Amtsblatt der Europäischen Union, L 327/1, 22.12.2000.

[3] Geiger, W.F. et al. (2005): Overall-effective measures for sustainable water resources management in the coastal areas of Shandong Province, China in: Water Conservation and Management in Coastal Area (WCMCA), 13-15 November 2005, Qingdao, China.

[4] Holm-Mueller (1991): Die Nachfrage nach Umweltqualität in der Bundesrepublik Deutschland (= Umweltbundesamt, Berichte 4/91), Berlin.

[5] Londong, J.; Geiger, W. F.; Karl, H.; Meyer, P.; Meusel, S.; Hecht, D.; Werbeck, N. (2005): Auswahl von kosteneffizienten Maßnahmenkombinationen im Rahmen der Bewirtschaftungsplanung zur Erfüllung der

Financing Possibilities for Environmental Protection Measures

G. Würzberg

Regierungsbaumeister Schlegel, Munich, Germany

The People's Republic of China have performed a stunning development in the last 20 years. Large parts of the country have passed through a dramatic process of industrialisation. The People's Republic of China has succeeded in overcoming destitution and illiteracy. It is intelligible and comprehensible that the attention during the process of industrialisation was focussed particularly or nearly exclusive on the positive targets, like the supply of the population and the successful co-operation in the world market. But the inevitable consequence of such development is the minimisation of certain resources and the need of their protection with a careful and accurate cultivation and rationing in the future.

Worldwide the most fundamental resource of all recourses – the air – has recently become part of a methodical treating. According to the Kyoto-Protocol the trading with international emission rates has commenced since the beginning of this year. This means that every country has to adapt its carbon emission to the natural reduction of carbon or otherwise to buy emission-rates from countries in which area is an overage of reduction. The People's Republic of China and the United States of America haven't joined the climate agreement until now. But it has been noticed that the utilisation of the most fundamental resource has become object of a worldwide discussion.

After the air water ranks second as fundamental resource. Global targets are existing for water, too. In the *Millennium Development Goals* of the United Nations you can find one which includes handling with water. The target is: *Halve by 2015 the proportion of people without sustainable access to safe drinking water and basic sanitation.*

In the parts of the People's Republic of China, which don't participate in the new prosperity it is still necessary to improve the supply with water. But with the industrialisation new tasks have come along. The consumption of the resource water is often higher than the naturally regeneration in quantity and/or quality. Sinking groundwater levels and loading with pollution and pathogens are often the consequences.

In the area surrounding Qingdao the groundwater flow has already conversed because of the low groundwater level. The effect is, that saltwater from the ocean now penetrates into the groundwater, will make it inedible in the future.

A sufficient quantity of clean water is an integral part of quality of life and health, as well as a requirement for the further economic development. Particularly in industrial developed areas it is imperative to protect the resource water in an adequate way. Regularly with the realisation of the necessity of measures the question of financing simultaneously arises.

1 Financing ways

Basically there are the following kinds of financing for such projects.
— International Development Banks
— National Programs and Fundings
— Equity and Commercial Banks
— Investors, which can be national or international

This ranking can also be seen as a temporal sequence, whereas not all projects are appropriate to be financed in the last-mentioned two kinds of financing. Financing with commercial institutes or investors assumes a reflux of capital. therefore it is appropriate only for measures, which can be refunded by a tariff rating system.

Worldwide there are many international development banks giving support to build-up infrastructure, including the utilisation of potable water and environmental protection measures. I focus on two banks in this paper. First it is the most famous and worldwide largest development bank, the World bank, and then the most important German development bank, the "Kreditanstalt für Wiederaufbau", in short KfW, what means *Reconstruction Loan Corporation*. Capital which is given by the German Federal Ministry for Economic Cooperation and Development is usually also released by this bank.

2 World bank

The World Bank, which is owned by the countries of the United Nations, so as the People's Republic of China, is engaged in countries with very low economic status, in developing countries and in emerging nations, too. The last-mentioned are countries whose economics is in a process of development and consolidation. The People's Republic of China has an inconsistent form of appearance in this connection. While the areas in the Northwest are still very poor there are areas in the South and East with a high degree of industrialisation, up to cities like Shanghai, which could be mentioned as new prosperity centres.

The People's Republic of China has been working together with the World Bank since many years. At the World Bank's website 100 projects can be retraced until 1996.

How does a World Bank project arise?

Before a World Bank project comes into being a Country Assistance Strategy (CAS) has to be made. The Country Assistance Strategy is the World Bank's central framework for designing assistance programs for borrowers from the International Development Association (IDA) and the International Bank for Reconstruction and Development (IBRD = World Bank). A Country Assistance Strategy is developed for each country the bank works with. It identifies the key areas where the banks assistance can have the biggest impact on poverty reduction. A Country Assistance Strategy will take into account the priorities of the country's government and key stakeholders, the performance of its portfolio in the country and its creditworthiness. Issues such as what causes poverty, the characteristics of the poor, the state of institutional development, implementation capacity and governance are also factored in the Country Assistance Strategy. From this assessment, the level and composition of financial and technical assistance that the banks seeks to provide to the country is determined. The assistance strategies are designed to promote collaboration and coordination with the Banks partners.

Projects which are financed by the World Bank are arranged on a governmental level. The finance ministers of the member countries are connected with the task managers of the World Bank. Only The central governments of the countries can initiate projects, province governments or city parliaments cannot generate projects at the World Bank directly.

Usually a World Bank project contents a multitude of particular projects. After generating a project it becomes split up into smaller projects, international advertised for bids and assigned. During the process of generating the several sub-projects, province governments, municipalities and associations absolutely have many possibilities of exertion of influence. But also concerning the particular projects, any decision has to be made by the central government and the task manager of the World Bank.

The capital which is given by the World Bank is either grant benefit which needn't paid back or loan.

Projects of the World Bank usually are large and comprehensive. Therefore they take long cycles. Due to the need that generating of the projects has to be done by the central government according to the Country Assistance Strategy, the need of time of the prearrangement can extend to several years. If a Project already exists, it is more easy, to define a project as part of the whole one.

3 Green development bank

The German Reconstruction Loan Corporation has given considerable capital to the People's Republic of China in the last decade. Most years the People's Republic of China was beyond the first five countries as measured by the height of the amounts given by the bank. Apart from a few exceptions the sum of the annual financing was from 100 millions € to 350 millions €. The Reconstruction Loan Corporation defined a focal point of promotion for the People's Republic of China. Especially wastewater and waste treatment are pronounced in this context.

The financial co-operation of the Reconstruction Loan Corporation begins when huge respectively heavy polluted waste water couldn't be coped with easy methods and advanced technologies are used. Planning, equipment and education of the personnel are financed. Condition precedent to the promotion is the implementation of cost-covering rates.

The Know-How-Transfer of line-operations becomes more and more important. In case of Public-Private-Partnership projects for example, Chinese enterprises should co-operate with international experienced operating companies.

The projects don't have the dimension like the World Bank ones in the respect of financing and time. But there should be calculated a period of minimum 6 month for the prearrangement. The Reconstruction Loan Corporation isn't predisposed to agree projects on the governmental level.

The financial instruments of the German Reconstruction Loan Corporation are follows:
— composite finance
— promotional loans
— mixed financing
— loans at reduced interest rates

3.1 Composite tinance

Composite Loans are available for financing projects that are eligible for promotion according to development-policy criteria in the fields of infrastructure, e.g., water supply and waste management and environmental technology. Borrowers may be states or project-executing agencies benefiting from a state guarantee.

The terms of the Composite Loans are adapted to the project periods. The interest rate for such a Composite Loan is either variable or fixed for the entire loan term. The Composite Loan interest rate is always below the market interest level. By adapting this instrument to the economic situation of the partner countries and the commercial viability of the projects KfW is able to provide financing solutions tailored to individual needs.

Composite Loans may either be tied or untied to German supplies. Untied Composite Loans must contain a grant element of at least 25% to qualify as official development aid.

Tied Composite Loans are available only for projects that are commercially non-viable owing to the OECD Consensus rules. In other words, they can be granted only for projects with cash flows insufficient under commercial financing conditions. The grant element required in this case is at least 35%.

3.2 Promotional loan

The promotional loan is an instrument that has been designed to finance development projects and primarily in countries without apparent indebtedness problems. With terms and conditions that are similar to those offered on the capital market, it rounds off the range of available instruments. It closes the gap between development loans and commercial financing schemes. Promotional loans complement the contributions of German development cooperation and are particularly appropriate for promoting the private sector in developing countries, which is increasingly gaining in importance.

Financing may be provided for private and public investments in infrastructure for example water supply.

The award of these funds is conditional in any case on the developmental soundness (=Kreditwürdigkeit) of the projects to be financed. Their appraisal is performed on the basis of the sectoral and regional development policy principles of the German federal government. In addition, the project risks must be acceptable to KfW and the borrower's credit worthiness must be satisfactory. Promotional loans are granted at conditions that are similar to those of the market.

The forms of financing depend on the creditworthiness of the borrower/project-executing agency and its specific needs. In this regard KfW is able to make a flexible offer of maturities, currencies and interest rate options that fully meets the needs of the customer. In the area of infrastructure finance promotional loans can be granted to state borrowers, private enterprises or within the framework of project financing schemes.

In all cases the loans are untied to supplies.

Particularly in project financings in the area of economic infrastructure KfW should be contacted as early as possible to accommodate the complex structure of this financing form. An information memorandum prepared by the sponsors/investors may already be the basis for a preliminary risk assessment. The detailed credit risk analysis, however, is performed on the basis of a feasibility study.

Together with the project participants a viable financing concept, including collateral, will then be worked out.

3.3 Mixed financings

In addition to the budget funds made available by the German federal government, KfW extends capital market funds at favourable conditions. This increases the volume of German Financial Cooperation. By adapting this instrument to the economic situation of the partner countries and the commercial viability of the projects KfW is able to provide financing solutions tailored to individual needs.

Borrowers may be states or project-executing agencies benefiting from a state guarantee.

Under mixed finance, KfW complements the budget funds by capital market funds. The grant element of the total financing meets the requirements for recognition as official development aid (ODA). The country risk is assumed by KfW, which therefore requires this risk to be covered by a Hermes coverage (supply-tied fixed finance) or by a first-class foreign official export credit agency (untied mixed financing). Consequently, mixed finance projects also have to meet the requirements of the respective export credit insurance agency. The fees for the export credit insurance must be considered in the supplier's offers and must be borne by the project-executing agency.

The conditions for mixed financings essentially depend on the conditions at which the budget funds are extended to the partner country. Particularly needy developing countries with a per-capita income of up to USD 875 are eligible for loans at IDA conditions (0.75% p. a. interest, 10 grace years, term of 40 years). All other developing countries may obtain these loans at standard FC conditions (2% p. a. interest, 10 grace years, term of 30 years). These budget funds may be combined with a hypothetical loan of market funds

3.4 Reduced-interest loan

The reduced-interest loan, is also available to the developing countries. This facility provides additional capital to advanced developing. By adapting this instrument to the economic situation of the partner countries and the commercial viability of the projects KfW is able to provide financing solutions tailored to individuals needs.

Given the relatively short maturity of the loan, this instrument particularly serves the financial sector. In principle, it is also available for financing projects that are eligible for promotion according to development policy criteria in the fields of infrastructure (telecommunications, energy generation and distribution, transport, environmental technology and waste management) and industry. Reduced-interest loans are not tied to German supplies. KfW uses only capital market funds for the reduced-interest loans. Interest payments under the loan are reduced by means of grant funds from the federal budget so that the requirements are met for recognition as official development aid (ODA). The grant amount is set according to the market interest rate level and the exact amount can be determined only at the time of conclusion of the contract.

Reduced-interest loans may be borrowed by States or project-executing agencies benefiting from a state guarantee.

The maturity of the reduced-interest loan is generally 10 to 12 years with 2 to 3 grace years. Since the reduced-interest loans are granted at KfW's own risk, they are awarded only if KfW considers the risks acceptable. Depending on the conditions agreed, the interest rate to be paid by the borrower currently ranges between 4.25% p. a. (maturity of 10 years and two grace years) and 5.00% p.a. (maturity of 12 years and three grace years).

4 National Funding

In the rate as financial means of international development banks are no more sufficient, national efforts are necessary. Indeed international promotion can go along with substantial progress, but the final solution of a problem has to come out of the country itself.

The tasks arising with the protection and the care of the resource water and its distribution, as well as the flood protection can be divided in 2 basic categories. On the one hand there are measures which are in superordinate public interest and have to be paid by public funding. On the other hand there are measures which can be refunded by tariff rates.

The superordinate public tasks primarily are the legislation and the creation of administrative structures to organize the utilisation and regeneration of water. The corresponding effort regarding the costs and required expertise partly can be supported by international aid. But on the whole this is a task of the country itself. This also has to be considered for the case of flood control.

But the costs arising with the supply with potable water or industrial process water, as well as the waste water disposal can completely be borne by the users and beneficiaries, if there is a certain degree of industrialization. Pre-condition is the existence of a tariff rating system. As I know this is already fact in most parts of the China. A tariff rating system also protects against abuse. I want to tell you a personal experience.

About 20 years ago there was a political change in the eastern Part of Germany. The former German Democratic Republic joined to the Federal Republic of Germany. Thereon market-economy has been introduced in the former German Democratic Republic. In midsummer I had a talk with an employee of an eastern German water supplier. He said, since the introduction of market economy there is a sufficient quantity of water all over the year, also in dry periods. I asked him, whether they had installed new pumps. No, but they had introduced a new tariff rating system and meters for every user. Since then they had no more lack of water.

The reason of the shortage of water was an useless consumption which took place in the allotments (=Schrebergärten). Every allotment holder (=Schrebergärtner) had a rain barrel (Regentonne) with a garden hose inside. In spring they opened the water tap a little and let it open through the whole year until they closed it again before winter.

A tariff rating system is a really useful feature to limit the usage of water to the necessary quantity and to provide its funding as well. A tariff rating system opens the possibility to finance measures in the field of water supply and waste water disposal with commercial banks. In combination with a sustainable solid legislation also international investors will be interested. In the field of water supply this is already Reality in many places in China. Many foreign companies have already invested in BOT-projects.

In principle it is also possible to run waste water treatment plants as BOT-projects. Precondition is a consistent information of the population, which in general has no sufficient willingness to pay for this.

In Germany the facilities of water supply and waste water disposal are nearly complete in public ownership, nevertheless they are organized as commercial companies and working profitably. The advantage of this model is, that the municipalities don't lose control over the valuable water resources and they are flexible to adapt the quality of waste water treatment to changing requirements.

5 Summary

It is always attractive to finance necessary measures with international advancement. If you attempt a financing by the World Bank you should proof if there is already a project in which the measure could be integrated. Otherwise it would be necessary to generate a new project in accordance with the Country Assistance Strategy over the central government. Therefore a period of a few years has to be calculated. World Bank projects have to be decided in agreement with the central government.

Co-operation with this bank (German Reconstruction Loan Corporation) especially is interesting if the project isn't as large as a World Bank project and shouldn't take so much time. Normally the participation is not tied to certain supplies. But normally together with the financial co-operation with the KfW, an export of German technology and know-how takes place. Negotiations with the German Reconstruction Loan Corporation can be made by local authorities also.

In principle such benefits serve the purpose to promote the country to accomplish this problems on their own in the future. To pay for the financial expenditure – we spoke about treating water resources – we advise to set up tariff rating systems in the range of drinking-water management and waste water management to assure the refinance of the necessary input.

You can charge national and international companies with the task of water supply and waste water treatment, but also you can run municipal companies with the same effect. In Germany municipal companies are prevailing and provide an excellent water management and water disposal.

WBalMo: Water Resources Planning Model
– Introduction and Practical Experience

S. O. Kaden
WASY Institute for Water Resources Planning and
Systems Research Ltd. Berlin, Germany

1 Water resources planning – stochastic problems

Water management is integrated into the hydrological cycle, which is shaped by precipitation and evaporation as the dominant factors in runoff generation, and runoff into the surface waters and into groundwater.

Superimposed on these natural processes are anthropogenic components, such as use of surface water and groundwater, and water-management measures such as reservoir control or transfers. In mining regions, as it is typically for the Spree river basin, discussed in this paper, both, the discharge of mine drainage and the remaining pits have to be considered too.

The goals of water management are, for example, covering the water needs of the users (as municipal water supply, power plants, industry), maintaining minimum discharges (for ecol- ogy reasons but also for navigation), and effective protection against floods.

Water management is a stochastic problem. The drainage process itself is a deterministic one, but lack of knowledge of the hydro-meteorological processes driving it, and of the spatial- temporal distribution of runoff generation, forces us to regard the runoff process over the long term as a random process. Figure 1 illustrates the two alternatives for generating time series of runoff. For the simulation and analysis of long term water resources planning, usually monthly time steps are used.

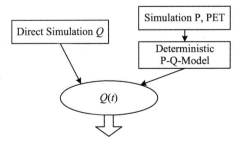

Figure 1   Runoff generation

Obviously, by employing methods of adequate stochastic simulation models one can generate different scenarios of precipitation and resulting runoff time series for a certain period of time.

A second aspect of uncertainty is related to possible climatic changes in the future.

The user demands, on the other hand, are deterministic in time and space from the planner's point of view, although they may well depend on meteorological variables, for example. Nevertheless user demands in future are dependent on uncertain socio-economic development.

Finally water resources planning results in management measures as the construction or operation of reservoirs, depending on water resources availability and user demand in future. Figure 2 summarizes the aspects discussed above.

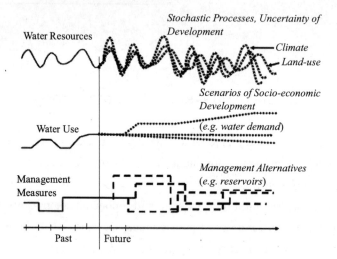

Figure 2   Aspects of water resources planning

Based on the stochastic character of runoff, and thus of water management, the methodology of stochastic long-term management has been developed, mainly for areas characterized by a large demand for water and small water resources available. In the Eastern part of Germany this is especially the mining area of Lower Lusatia in the Spree river basin.

The stochastic management models divide the management problem into three parts (Figure 3):

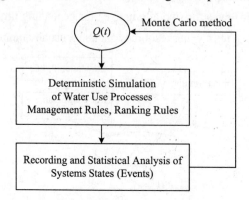

Figure 3   Method of stochastic water resources management modeling

- Stochastic simulation of meteorological and hydrological processes
- Deterministic simulation of water use processes
- Recording of relevant system states

If the month-by-month simulation is done over sufficiently long periods, a statistical analysis of the recorded system states will give satisfactory approximations to the probability distribu- tions sought, for reservoir levels and discharges for certain water-balance profiles, or safety margins for water provision, for example.

2 The simulation system WBalMo

Based on the general methodology described above a series of models, the so-called long-term planning models has been developed in the last decades (Schramm 1979). The first generalized model – GRM - of this type was developed in the 80s of the last century (Kozersky 1981). In the 90s the model was further improved by the WASY Institute for Water Resources Planning and Systems Research Ltd. and adapted to new hard- and software technologies. The recent technology is the WBalMo model system, with an ArcView user interface.

Based on the knowledge of a river basins structure as well as specific natural runoff and water resource processes the quantitative behavior of a river basins water resource system can, un- der various conditions, be examined by the WBalMo software system.

The management model is based on the Monte-Carlo method which forms the reclining basis of the WBalMo software system. It allows a river basins water utilization processes to be re- produced covering intervals of 1 month for any time period. Parallel to this the registration of relevant system affairs facilitates a statistical analysis of registered events on completion of the simulation. As a result approximate probability of distribution for areas, such as reservoir capacity, water supply deficiency for individual water users or for flow, are available at se- lected river profiles. Due to this, the quality of a selected management strategy can be as- sessed for the investigated river basin and a gradual improvement of this strategy can then be achieved with aimed variant calculations.

Treatment of stochastic entry sizes and deterministic reproduction of water utilization proc- esses are strictly separated in WBalMo. For the chosen time period, on the basis of a month, WBalMo calculates the dependant temporal distribution of runoff generation by producing a chronological series of entry sizes. By rule this is evaluated via a stochastic simulation model under compliance with time dependant conditions for drainage formation.

The procedure for one monthly time step is shown in Figure 4.

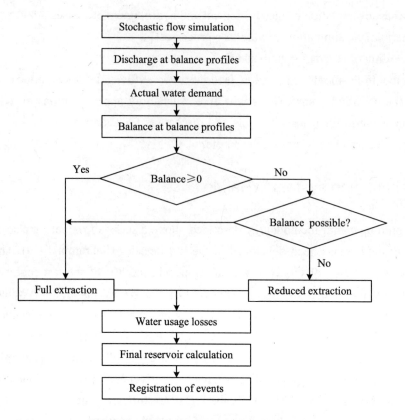

Figure 4 Balancing procedure of WBalMo

Model assumptions

The following assumptions form the reproduction basis of the water utilization processes within the model system WBalMo:
- WBalMo is based on a schematic representation of hydrological processes in a river basin by the use of Flowing Waters and Balance Profiles.
- By subdividing the whole flooded area, Simulation Sub Areas are worked out and assigned to the above named discharge time series. This discharge is then shared out between balance profiles as separate availability.
- Consideration of User water utilizations takes place in accordance with their location and size, which extracts and feeds back the necessary water from balance profiles.
- WBalMo is based on inclusion of Reservoirs (surface water reservoirs, lakes) by details of their location, capacity and Output Elements, which describe their active orientated requirement.

An example of the WBalMo systems structure is shown in Figure 5.

Ranking of users

All users and outputs are allocated a *number rank*. This enables every user and their purpose to be filed into the whole system (e.g. drinking water supply for agricultural irrigation). The reservoir output availability for different users or user groups is subject to change, for example significant use during dry periods due to reservoir outputs is still possible, whilst other periods have to be prepared to be reduced. Modeling of user processes takes place according to a fixed algorithm. "Dependency on users" corresponding to the ranking list results in different requirements for reservoir releases.

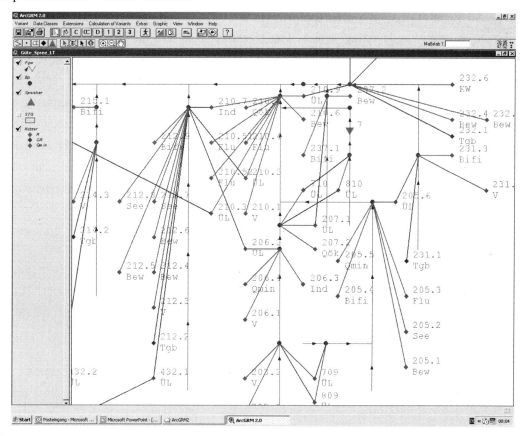

Figure 5   Model structure of WBalMo

Innovations in the recent WBalMo version are:
- system sketch visualization of the river basin within the program system
- filling of model data into a data base and
- availability of an interface for external models, which allows special investigations (e.g. water quality, daily simulation of values) on the basis of the WBalMo quantity model

3   Water resources planning for the spree river basin

The river basin

The Spree river basin with a catchment area of about 10.000 km² is located in the South-

Eastern part of Germany. A rough overview is given in Figure 6. The hot spots are the Lusatian lignite mining district, the Spreewald (an important wetland region) and the city of Berlin.

Problems of water management in the lignite mining regions

The Lusatian lignite mining district is located about 80 km south-east from Berlin. As a result of the one-sided energy politics of the former GDR German Democratic Republic the brown coal (lignite) deposits were exploited without consideration of ecological aspects. The artifi- cial groundwater lowering, which was necessary to enable the brown coal quarrying, affected an area of almost 2 100 km^2. The drainage water was pumped into rivers, in particular the River Spree.

The discharge into the receiving streams was strongly increased. As a consequence the drainage water diluted municipal and industrial wastewater which, therefore, had only a small influence on water quality. However, the influence of the mining induced parameters was heavily increased. By oxidation of pyrite, which is recognized as the major source of acid mine drainage, high concentrations of iron and sulfate are brought into the receiving streams (Schlaeger et al., 2000).

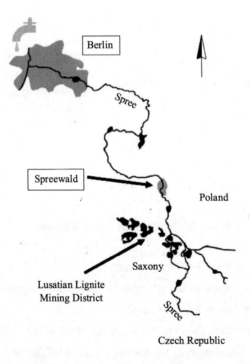

Figure 6   Spree river basin, rough overview

After the German reunion in 1990 the exploitation of most of the open-cast mining was stopped because of economical aspects. Due to this the discharges of the River Spree decreased. Moreover, water taken from the River Spree and its tributaries is now used to flood the remaining pits from open-cast mining as well as to refill the draw-down cones. The resulting water shortage affects the minimum flow for the Spreewald wetland, the water supply in Berlin etc..

As a result of the occurring water shortage the lack of dilution also influences water quality considerably high. In addition, the mining induced parameters continue to have an impact on water quality caused by the arising groundwater level and ongoing discharge of mine drainage.

WBalMo Spree – Schwarze Elster

In order to model the highly complicated and interrelated processes of runoff formation and water use in the river basin (together with the basin of Schwarze Elster), an WBalMo management model has been developed and is permanently used by the responsible water authorities.

The complexity of the model may be illustrated by the following model parameters:
- 170 balance profiles
- 64 simulation sub areas
- 415 water user
- 14 reservoirs and 52 reservoir releases
- 53 DYN-elements (programmed sub-routines for special, non-standard sub-processes)
- 280 statistical registrations

One of the applications of the model is the analysis of the need and the consequences of the construction of new reservoirs, or of water transfer from other regions (the Odra river basin). As reservoirs in the mining region large remaining pits might be used. The next Figure 7 illus- trates model results for different scenarios.

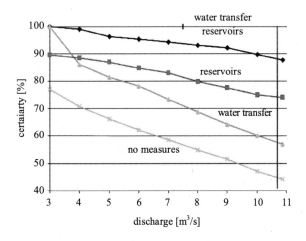

Figure 7 Scenarios for the discharge of River Spree, gauge Große Tränke

The x-axis is the discharge in the River Spree at the gauge Große Tränke (near Berlin). The required minimum flow amounts to about 11 m^3/s. The y-axis is the certainty to guarantee a certain discharge. The figure shows that without measures every other year the required discharge is not guaranteed. Almost every other year the discharge amounts to only 50% of the required one. The effects of different measures to stabilize the discharge are obvious. But even with reservoirs and water transfer, the required discharge cannot be guaranteed ap- proximately every 10 years.

4 Actual new developments

Modularization of WBalMo

In the case of the Spree river basin there are two important sub-regions, which have been developed separately – the Berlin region and the Spreewald wetland. The Berlin module has been included into the WBalMo completely with all details. A different situation is given for the Spreewald wetland. For detailed studies of the wetland system a separate Spreewald model had to be developed. This model has not been included into the WBalMo Spree - Schwarze Elster, but it is directly coupled. It is of course not possible to see details in the, but it illustrates the high complexity of the modeled water systems with the river system, users, reservoirs and balance points.

In case of still larger river basins with important sub-basins the aspect of modularization is even more important, because a single management model may become rather large and not easy to handle. Furthermore, frequently models for sub-basins already exist, which have to be embedded in a large model later. The coupling between the models has to be interactive, i. e. that management decisions within one sub-model may effect others. A vision for the development of such a modular WBalMo for the large Elbe river basin (about 150 000 km^2) is depicted in Figure 8.

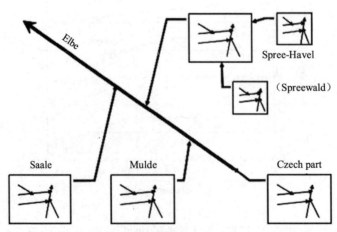

Figure 8   Concept WBalMo Elbe

Long-term planning and water quality

As discussed above water quality is an important control value in the Spree river basin, too. Besides non-point pollution originating mainly from agriculture and point pollution from industry and municipalities the water quality impact of existing and closed lignite mines is most important. The significant transformation processes between these parameters as well as interactions between the parameters and sediment or atmosphere has to be considered.

A first WBalMo-Quality for the upper Spree river basin has been developed by Schlaeger et al. (2000). To capture all relevant intakes for water quality and to describe the water balance at the sampling stations the base system structure had to be enhanced. Additional balance pro- files as well as new water users (sewage plants, industry, fishing ponds, etc.) had to be im- plemented. Figure 9 shows the relevant parameters.

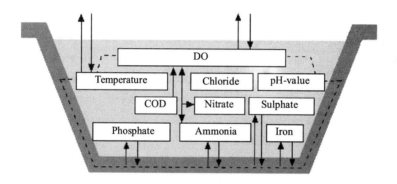

Figure 9   Relevant water quality parameters in lignite mining regions

The water quality model is linked to the water balance model by the delivery of the simulated discharges Q_i at balance profiles, the water usage, lateral, non-point discharges and the actual time step or actual month, respectively. Time of flow t, which is important for simulating the transformation processes, is calculated by characteristic functions $f(Q_i)$, which are determined for defined river sections.

Since the simulation of water quality in a long-term balance model is subject to a rougher consideration than in a process model, the elaborated transformation processes have to be simplified but still have to describe the essential reactions sufficiently precise. In Schlaeger et al. (2000) the method for developing such reduced water quality models is described.

As an example of modeling results Figure 10 shows the modeled influence of discharged mine drainage on COD.

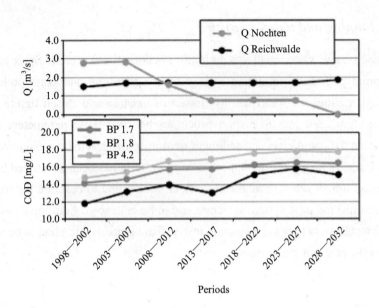

Figure 10 Influence of discharged mine drainage on COD (chemical oxygen demand)

The development of this model was a pilot project for a project in which the entire Spree catchment area (from its spring region up to Berlin) is considered.

References

[1] Kozerski, D. (1981): *Rechenprogrammsystem GRM als verallgemeinertes Langfristbewirtschaftungsmodell*. In: Wasserwirtschaft-Wassertechnik 31, H. 11/12, S. 380-394, 415-419.

[2] Schlaeger, F., Köngeter, J., Redetzky, M., Kaden, S. (2000): *Modeling and Forecasting for Water Resources Management: The River Spree Project* Paper at Hydroinformatics 2000, Iowa, USA, July 2000.

[3] Schramm, M. (1979): Zur Anwendung stochastischer Methoden bei der Modellierung wasserwirtschaftlicher *Systeme*. Dresden (Habilitationsschrift, TU Dresden).